IET POWER AND ENERGY SERIES 53

T0258378

Condition Assessment of High Voltage Insulation in Power System Equipment

Other volumes in this series:

Condition Assessment of High Voltage Insulation in Power System Equipment

R.E. James and Q. Su

The Institution of Engineering and Technology

Published by The Institution of Engineering and Technology, London, United Kingdom

© 2008 The Institution of Engineering and Technology

First published 2008

The Institution of Engineering and Technology
Michael Faraday House
Six Hills Way, Stevenage
Herts, SG1 2AY, United Kingdom

www.theiet.org

British Library Cataloguing in Publication Data
James, R. E.
    Condition assessment of high voltage insulation in power
    system equipment. - (Power & energy series; v. 53)
    1. Electric insulators and insulation - Testing 2. High Voltages
    I. Title II. Su, Q. III. Institution of Engineering and Technology
    621.3'1937

ISBN 978-0-86341-737-5

Typeset in India by Newgen Imaging Systems (P) Ltd, Chennai
Printed in the UK by Athenaeum Press Ltd, Gateshead, Tyne & Wear

Contents

Both authors have significant industrial experience in the United Kingdom (REJ) and China/Singapore (QS) and in teaching and research at Portsmouth Polytechnic (REJ), University of NSW (REJ, QS) and Monash University (QS). During the latter periods many consultancies concerned with industrial high-voltage insulation problems were undertaken. We wish to acknowledge the value of our association with many ex-colleagues in industry and the universities and those in the various utilities with whom we have worked.

Our thanks are especially due to our wives and families – Felicia (REJ), Liling and daughter Shirley (QS) – for their patience and understanding during writing of the book.

R.E. James
Q. Su

Chapter 1

Introduction

- Power system components
- Insulation coordination concepts and high-voltage test levels
- Need for insulation condition monitoring

A high-voltage power system consists of a complex configuration of generators, long-distance transmission lines and localized distribution networks with above- and below-ground conductors for delivering energy to users. This introductory chapter indicates the wide range of high-voltage components whose successful operation depends on the correct choice of the electrical insulation for the particular application and voltage level. The condition of the insulating materials when new, and especially as they age, is a critical factor in determining the life of much equipment. The need for effective maintenance, including continuous insulation monitoring in many cases, is becoming an important requirement in the asset management of existing and planned power systems.

As the voltages and powers to be transmitted increased over the past hundred years the basic dielectrics greatly improved following extensive research by industry and in specialized laboratories, where much of this work continues. It is of interest to note that paper, suitably dried and impregnated, is still used for many high-voltage applications. New dielectrics are being introduced based on many years of research and development and are becoming more widespread as operational experience is obtained. In order to ensure an economic power-supply system with a high level of reliability, it is important to be able to monitor the dielectric parameters of the various insulations being utilized – when new and in service. Later chapters describe the materials and their applications, including examples of possible fault scenarios, dielectric testing techniques for completed equipment, new and existing condition monitoring systems and, finally, the application of artificial intelligence in incipient fault diagnosis and condition assessment.

Present power systems are ageing significantly and in many cases 40 per cent of the equipment is older than the conventional 'design life' of 25 years. This figure was probably chosen because of the uncertainties in estimating the anticipated lives of the practical insulation structures and for commercial reasons. In fact, many system components are still functioning satisfactorily after much longer periods. This is possibly due to the relatively low average electric stress values used to allow for inherent inaccuracies in calculations of maximum values within the complex structures. The development of suitable computer programs has enabled much improved designs to be achieved. Also, in many systems the circuits were operated in parallel to cater for overloading and possible failure of one line or unit. This configuration probably resulted in the average dielectric temperatures being below the allowable maxima. The situation is changing with the need for the managed assets to realize maximum economic returns. It is only by effective condition monitoring over long periods that data can be acquired, thus enabling the rate of deterioration of the insulation structures to be determined in service. This would naturally include the influence of possible generic manufacturing and design faults as well as inappropriate maintenance. Trends in such data assist in the more reliable prediction of the remaining life of equipment, possibly including the application of probabilistic techniques.

1.1   Interconnection of HV power system components

Contemporary system voltages range up to 1 000 kV(RMS three-phase) or higher and 600 kV(DC), although the more usual AC values are 500/750 kV and below. Bulk powers greater than 1 000 MW may be transmitted by a single three-phase circuit over long distances, in some cases for more than several hundred kilometres. Local delivery ratings may be of many tens of MVA down to a few kVA.

The application of renewable sources – for example solar devices, wind generators, biomass generation and small hydro-plants – is becoming more important. Within ten years it might be expected that embedded generation from such sources could contribute between 10 and 20 per cent of the total power in some countries, although commercial problems may limit the developments [1]. The form of the existing power system infrastructures would probably not change significantly for such conditions, especially where high levels of energy are required at a particular location. The newer sources will operate locally at low voltages and include conventional step-up systems where they are coupled to the main distribution/transmission system. Special insulation problems will be involved but these are outside the scope of the conditions considered in this book. Descriptions of how renewable sources are being developed and the possible effects of their dispersion within the established power systems are discussed in the literature.

Although the majority of power systems transmit at alternating frequencies, a significant number incorporate direct voltages. This requires special equipment and introduces different insulation problems, some of which are considered later.

1.1.1   *Alternating voltage systems*

The major components of a system with their possible relative locations at a power station and in substations are indicated simply in Figure 1.1.

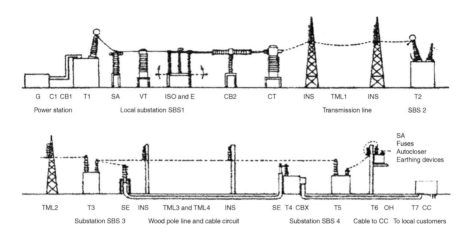

G	C1 CB1	T1	SA	VT	ISO and E	CB2	CT	INS	TML1	INS	T2
Power station			Local substation SBS1						Transmission line		SBS 2

SA
Fuses
Autocloser
Earthing devices

TML2	T3	SE	INS	TML3 and TML4	INS	SE	T4	CBX	T5	T6	OH	T7	CC
	Substation SBS 3			Wood pole line and cable circuit			Substation SBS 4			Cable to CC		To local customers	

Figure 1.1   Basic system for generation, transmission and distribution of AC power

The systems are based on three-phase configurations, although many of the individual elements are single-phase. Each device must have appropriate electrical insulation for its particular structure. Many of the methods by which this is achieved are discussed in Chapters 3–5 and techniques for assessing the condition of the materials when new and in service are described in subsequent chapters.

At the power stations the generators (**G**) may be driven by diesel (oil) engines, gas turbines, water turbines or steam turbines – the last of these being most usual for the larger machines. Generation voltages in large systems range from 12 kV to 24 kV (perhaps up to 33 kV in a few cases) with current ratings of 1 500 A up to 16,000 A or larger. These high currents are fed through cables (**C1**), or metal-enclosed bus conductors of large cross sections, to the low-voltage windings of the step-up 'generator' transformers (**T1**). High-current circuit breakers (**CB1**) may be installed between the generator and transformers. The conductors required from the high-voltage terminals of **T1** are of reduced dimensions, thus allowing power transfer by the use of bare overhead cables through a local substation (**SBS1**) and then over long distances (**TML1**) or, within cities, through fully insulated underground cables (**TML4**).

At the receiving end of the various lines, a step-down 'transmission' transformer (**T2**) is connected. Such units are often wound as autotransformers, especially if the lower voltage is at an intermediate level (e.g. 145 kV) for secondary transmission (**TML2**) or for supplying a city's major distribution system. The system feeds double-wound transformers (**T3**) with outputs of the order of 66–33 kV (**TML3 and TML4**) for reduction of the voltages (**T4–T7**) to customer operating levels in the range 12 kV to 415 V/220 V/110 V. A cable-fed control cubicle (**T7CC**) is shown for underground supply to a number of domestic customers. Large industrial organizations may purchase power at the higher voltages and install their own local substation. The choice of voltage ratios, and the required transformer impedance values between windings depend on many factors related to the particular supply and load conditions. Numerous books and technical papers have been published on this subject [2].

At the major changes in voltage where primary lines (or generator(s)) feed a number of other lines a substation (**SBS1–SBS4**) is constructed for control of the individual circuits: for monitoring the real and reactive power flows possibly including an optical fibre-coupled thyristor firing system for operation of static VAR compensators (perhaps of the relocatable form [3]) and for protection of the system when subjected to faults and overvoltages. The various devices, some of which are represented in Figure 1.1, must be insulated for the different service voltages – including surges – to ground and between phases. Switching and isolation are provided by circuit breakers and air isolators (**CB2** and **ISO**). The current magnitudes and steady state voltages are monitored by current (**CT**) and voltage (**VT**) transformers of various designs. Surge voltages due to lightning and switching are limited by surge arresters (**SA**) and air gaps – for example across transformer bushings (**T2**), circuit-breaker insulation and at the entry to a substation. Where a high-voltage conductor passes through an earthed tank a bushing is required as in power transformers, 'dead tank' instrument transformers, some older oil circuit breakers and in gas-insulated systems (GIS). The overhead lines (**TML1–TML3**) must be supported with insulator strings or similar (**INS**) capable of withstanding the various voltages and adverse weather conditions – again rod gaps and surge arresters may be utilized for protection.

The machine floors of a steam-turbine-generator and a hydro-generator power station are depicted in Figures 1.2 and 1.3 respectively. The complexity of outdoor substations is indicated in photographs, Figures 1.4–1.6, in which may be identified many of the items in Figure 1.1. The components in substation SBS1 are present in one form or other in all levels of high-voltage substations often including cable-sealing ends (SE) as in SBS3 and SBS4. A large system would involve many lines and plant items.

Following the development of GIS, it has been possible to design and build compact substations for very high voltages. Many examples of this application exist where space is limited – near or in major cities.

At the lower voltages much maintenance is necessary to ensure high reliability of supply in the local distribution system. In Figure 1.7 are shown different aspects of such a system in a built-up area. The 415 V house supplies are fed from the 11 kV overhead lines through a pole-mounted transformer, Figure 1.7(a), and Figure 1.1 (T6), by means of either overhead wires (Figure 1.7(a)) or a three-phase cable to cubicles located several hundred metres away in a housing complex (Figure 1.7(c)). Each unit (Figure 1.7(c) and T7CC in Figure 1.1) contains a step-down transformer with appropriate protection for the outgoing 415 V cable circuits. The items of particular interest in relation to insulation are the surge arresters, cable and sealing ends, 11 kV fuses, the various line and stand-off insulators, the oil-filled transformer and, of course, the wooden poles. The pole in Figure 1.7(b) was being replaced because of termite damage but this operation could have been necessary following a lightning strike or even a bush fire.

Figure 1.2   Steam turbine-generator [4] [reproduced by permission of CIGRE]

Figure 1.3   Hydro-generator [5] [reproduced by permission of CIGRE]

Figure 1.4   330 kV substation. Note (from left to right) – current transformers, SF$_6$ circuit breakers, support insulators for air isolators/automatic earthing arms [reproduced by permission of TRANSGRID, New South Wales]

Figure 1.5   330 kV substation. Note insulator strings and corona rings [reproduced by permission of TRANSGRID, New South Wales]

Figure 1.6   132 kV isolator with good corona design. Note automatic earthing arms in foreground

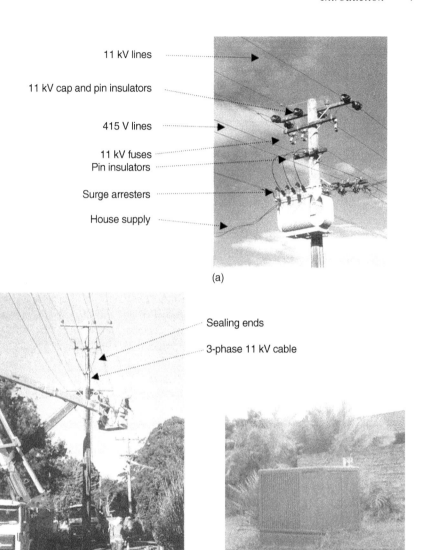

11 kV lines

11 kV cap and pin insulators

415 V lines

11 kV fuses
Pin insulators

Surge arresters

House supply

(a)

Sealing ends

3-phase 11 kV cable

(b)   (c)

Figure 1.7   11 kV and 415 V local supply systems. (a) 11 kV/415 V pole-mounted transformer; (b) 11 kV overhead line to 11 kV three-phase cable. Pole maintenance; (c) A 415 V cubicle substation.
Note surge arresters, 11 kV cable sealing ends, 11 kV fuses, line insulators and 415 V wires to houses.

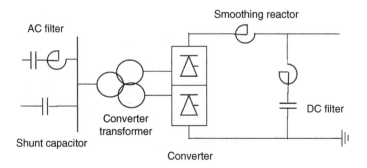

Figure 1.8   Principle of an HVDC transmission scheme

1.1.2   Direct-voltage systems

In effect a direct-voltage system is a hybrid circuit incorporating AC and DC components (see Figure 1.8). The incoming power is from an alternating source, which is rectified and filtered before transmission through the DC system, inversion taking place at the receiving end in order to provide the usual AC supply conditions. The harmonics produced by the converters are reduced by filters comprising R, L and C elements. The earlier significant systems included that from the Swedish mainland to Gotland (150 kV, 1954), the original cross-Channel connection between France and England ($\pm$ 100 kV, 1961), the crossing between the North and South Islands of New Zealand ($\pm$ 250 kV, 1965), the 50 Hz/60 Hz tie in Japan (125 kV, 1965), the link between Sardinia and the Italian mainland (200 kV, 1967), the overhead line from Volgograd to Donbass ($\pm$ 400 kV, 1965) and the Pacific Intertie in the USA ($\pm$ 400 kV, 1970).

These groundbreaking systems (and a few others) incorporated mercury-arc technology, which tended to reduce the attraction of HVDC transmission due to various operating problems. However, with the development of reliable high-voltage, high-power thyristors, the situation changed and there are now many systems worldwide. Such schemes are well established, transmitting 60 GW or more of the world's power [6]. Typical voltage levels, powers transmitted and line lengths, together with commissioning dates, are included in Table 1.1. The number of such schemes is probably approaching one hundred.

Modern systems use two 6-pulse bridges giving a 12-pulse converter bridge. One 'valve' consists of a number of thyristors – perhaps 100 series-connected units for 600 kV, each of which may be rated at about 8.5 kV maximum peak voltage withstand capability [7]. The number of thyristors required for 100 MW is quoted as 18 (compared with 234 thirty years ago) in Reference 8. The complexity of these structures has resulted in rigorous insulation testing procedures (see Section 7.11).

The advantages in respect of lower corona noise and losses, smaller wayleaves and the capability of being able to utilize cables for long lengths because of the reduction in losses compared with the three-phase AC equivalent may, in some applications,

Table 1.1   Examples of HVDC transmission schemes: thyristor valves

System	Voltage (kV)	Year	Capacity (MW)	OH line (km)	Cable (km)
Skagerrak	± 250	1976/77	500	113	127
Vancouver	± 280	1977/79	476	41	33
Nelson River BP2	± 250	1975/85	2 000	930	–
Hokkaido Honshu	± 250	1979/93	600	124	44
China	± 500	1987/98	1 200	1 100	–
Itaipu BP	± 600	1987	3 000	783	–
Cross Channel 2	± 270	1986	1 000	–	72
USSR	± 750	1985	1 500	2 400	–
Rihand–Dadri (Delhi)	± 500	1991	1 500	840	–
E–W Malaysia	± 350	1995	1 000	–	600
Garabi, Brazil	BtB Converters	2000/2002	2 200		
Three Gorges – Changzhou, China	± 500	2003	3 000		

offset the increased costs of converter stations compared with a corresponding AC system. These features will, of course, also be advantageous where environmental requirements are at a premium. The reliabilities of a significant number of the schemes are monitored regularly by WG 14.04 of SC 14 of CIGRE [9]: this report covers 28 thyristor valve and 5 mercury-arc valve systems operating during 1997/1998. Data were obtained initially in 1968.

Of special interest in respect of insulation assessment and possible monitoring are the converter transformers, which may be subjected to combined alternating and direct voltages, the smoothing reactors, the overhead line insulators, the bushings and especially any cables/accessories, particularly as used for underwater crossings.

With the new systems utilizing voltage-sourced converters (see Subsection 1.4.4) it appears that the insulation of equipment may be subjected to periodic impulse-type voltages [8], the effects of which have not been extensively investigated.

1.2   Insulation coordination

Insulation coordination design of power systems aims at minimizing outages of major items of plant and critical circuits caused by switching or lightning surges. The traditional protective methods use various forms of air gaps connected across particular equipment or transmission-line components. Because of the lack of matching between the V-T (volt-time) characteristics of the gaps and those of the non-restoring insulations in, for example, power transformers, the application may not be as effective as required. Also the gaps may allow the passage of high-value power frequency follow-through arcs.

Table 1.3   Shape of AC and impulse test voltages

Class	Shape	Frequency and time duration
AC tests		(1) $f = 50$ or $60\,Hz$ (2) VLF, e.g. $0.1\,Hz$ (3) Resonance voltage $(20-300\,Hz)$ (4) Induced voltage tests $(100/120-400/480\,Hz)$ (5) $T_t = 10$ seconds to 60 minutes
Switching-impulse tests		$T_p = 250\,\mu s$ $T_2 = 2500\,\mu s$
Lightning-impulse tests		$T_1 = 1.2\,\mu s$ $T_2 = 50\,\mu s$
Fast-impulse tests		$100\,ns \geq T_f > 3\,ns$ $0.3\,MHz < f_1 < 100\,MHz$ $30\,kHz < f_2 < 300\,kHz$ $T_t \leq 3\,ms$

high standard of maintenance and in-service monitoring is incorporated in the operational programmes, lives of 40 years are now being predicted. From Table 1.2 it will be seen that for values of $U_m = 300\,kV$ and above a short-time-duration power-frequency test is not specified, a switching surge test (see Subsection 1.3.3) being considered more appropriate. However, steady-state-test overvoltage proving tests are retained in some form or other. In the case of transformers this includes a partial discharge test (see Chapter 7) at, perhaps, $1.5 \times U_m\sqrt{3}$ for 60 minutes or more for $U_m \geq 300\,kV$ units. In many cases such tests are called for on lower-voltage equipment. The methods of producing the voltages for test purposes and procedures

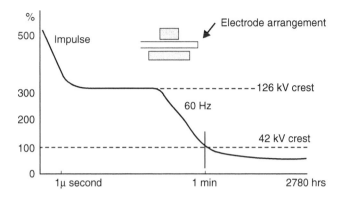

Figure 1.9   Volt-time curve for treated pressboard in oil – 1/16-inch-thick pressboard at 25°C [13]

for determining insulation properties are described in Chapter 6 and applications to particular apparatus in Chapter 7.

1.3.1   Power-frequency voltages

A power-frequency overvoltage test level of the same form as the operating voltage has been the traditional method for proving the integrity of equipment insulation as indicated in Table 1.3. An increase in frequency may be necessary for induced testing of transformers and in resonance-type tests as included in the table.

1.3.2   Lightning-impulse voltages

The need for tests to reduce failures due to lightning was appreciated very early in the development of transmission systems. This applied particularly to the transformers, as a power-frequency-induced test does not prove the integrity of complex inductive/capacitive structures when subjected to surges. The implications are analysed in Chapter 5, where it is shown that the behaviour under such conditions is a dominant feature in the insulation design of large transformer windings.

In the 1920s and 1930s it was realized that the reduction of the earthing (footing) resistance of the support towers of the overhead lines and the provision of overhead earth wires could give a moderate degree of protection. It was possible to measure lightning currents during a strike to the towers and to develop scenarios for determination of the magnitudes and forms of the surges to be expected. These include induction from near strikes, back flashovers due to high earth resistances and direct strikes to the power conductors. The various measurements, and basic observations, of strikes to lightning conductors – for example at the Empire State Building in New York and in South Africa – enabled a consensus to be established regarding the shape and magnitude of the stroke currents: in the range up to 150–300 kA with an average of perhaps 25 kA. Using this information in conjunction with calculated line-surge impedances, the shapes and magnitudes of expected voltages were estimated and agreement reached as to the form of 'standard' waves to be applied to equipment for particular voltage systems. The basic shape is 1.2/50 microseconds as defined in Table 1.3 and selected levels listed in Table 1.2. It will be noted that impulse-voltage magnitudes are of the order of 3.5 to 5 times the peak of the operating voltages to ground.

1.3.3 Switching surges

As the voltage levels of the transmission systems increased, it became apparent that the surges produced by switching could be more significant than the short-duration power-frequency overvoltages caused by faults and other operational abnormalities. The magnitudes of these surges are less than the lightning disturbances but are of slower rise-time and oscillatory in form. Thus, it was considered necessary to introduce a test to cover this condition. Although the surges are, in effect, oscillatory with frequencies in the range up to tens of kHz, this proved a difficult condition to simulate as a high-voltage test. Because an insulation failure or flashover – especially across air clearances in the transmission system – would be expected to occur during the first peak, it was agreed within the industry that a wave having a rise time of 250 microseconds and a decay of 2 500 microseconds would be representative and should be adopted as a standard test (see Table 1.3). Some of the recommended values are given in Table 1.2. The magnitudes are of the same order as the peaks of the short-duration power-frequency test voltages specified in previous versions of the standards.

1.3.4 Very fast transient tests (VFTT)

Following the extensive application of gas-insulated switchgear it became apparent that very fast transients were being injected into the power system during operation of disconnectors. These surges have rise times of tens of nanoseconds with superimposed oscillations in the range of tens of kHz up to 100 MHz as depicted in Table 1.3. Such disturbances can be dangerous if the gas-insulated switchgear equipment is directly connected to, for example, a transformer. The problems are under close scrutiny within CIGRE and elsewhere (see Chapter 8 of Reference 14).

1.3.5 Direct-voltage tests

Direct-voltage acceptance tests for HVDC equipment are of the order of two to three times the system operating values. The voltages are applied for long periods of, perhaps, 30 minutes to ensure the insulation system is in a stable, charged condition and the stress distribution across multiple dielectrics is identical to the service situation. Allowance must be made for systems where service polarity reversals are involved. Levels for converter transformers are included in IEC Standard 61378-2 [13] and a wide range of tests for the complex components of the thyristor valves in IEC Standard 60700.

Under certain conditions direct voltages are used for testing AC equipment, e.g. when withstand voltage testing of large rotating machines and cables where

4. Poidvin, D., 'Steam Turbine-Generator at Penly', *Important Achievements of CIGRE*, December 1998 (by permission of CIGRE)
5. Berenger, P., 'Hydro-Generator at Pouget', *Important Achievements of CIGRE*, December 1998 (by permission of CIGRE)
6. Asplund, G., 'HVDC using voltage source converters – a new way to build highly controllable and compact HVDC substations', CIGRE 2000, Paper P2-04 (on behalf of Study Committee 23)
7. Andersen, B., and Barker, C., 'A new era in HVDC?', *IEE Review*, March 2000, pp. 33–9
8. Lips, H.P., 'Voltage stresses and test requirements on equipment of HVDC converter stations and transmission cables', CIGRE 2000, Panel 2, Paper P2-06 (on behalf of Study Committee 14)
9. 'A survey of the reliability of HVDC systems throughout the world during 1997–1998', CIGRE 2000, Paper 14-02 (on behalf of Study Committee 14 – WG 14.04)
10. IEC 60071 Insulation Coordination Parts 1–3
11. IEC 60071 Insulation Coordination Part 5: 'Procedures for HVDC Converter Stations'
12. IEC 61378-2 'Converter Transformers' Part 2: Transformers for HVDC Applications
13. Montsinger, V.M., 'Breakdown curve for solid insulation', *Electrical Engineering*, December 1935;**54**:1300
14. Ryan, H.M. (ed.), *High Voltage Engineering and Testing*, 2nd edition (IET, UK, 2001)
15. Leijon, M., Dahlgren, M., Walfridsson, L., Ming, L. and Jaksts, A., 'A recent development in the electrical insulation systems of generators and transformers', *IEEE Electrical Insulation Magazine*, May/June 2001;**17**(3):10–15
16. 'Superconductivity makes its power transmission debut', general review in *Engineers Australia*, July 1999, pp. 26–30
17. Graham, J., Biledt, G., and Johansson, J., 'Power System Interconnections using HVDC Links', IX Symposium of Specialists in Electrical Operational and Expansion Planning (IX SEPOPE), 23–7 May 2004, Rio de Janeiro, Brazil, SP151
18. Brochure No. 269, CIGRE 'VSC Transmission', WG B4.37. See also Summary in *Electra* 219, April 2005, pp. 29–39

1.8 Problems

1. From personal observations and use of the Web, identify and record examples of the power system components mentioned in the chapter. Take advantage of visits to power stations and HV substations.
2. What is meant by 'embedded' generation and how is it utilized? Describe at least two such power sources and the methods for connecting them to the main high-voltage transmission system.

1.5 Future insulation monitoring requirements

The above survey of some of the complexities of a power supply system has high-lighted a number of components in which the maintenance of electrical insulation in good condition is essential in order to achieve efficient and safe operation. Such objectives can be realized at the engineering level only by the application of appropriate monitoring techniques, in particular those associated with assessing the state of the insulating materials in equipment when new and during lives of up to forty years or longer. It must be accepted that insulating materials inevitably deteriorate with time – the rate being very dependent on usage and the quality of maintenance achieved.

The choice of whether or not to incorporate simple or advanced monitoring instrumentation will depend on many factors including the replacement/repair cost of the particular equipment and, probably more importantly, the overall economic effects and associated disruption of the power system following a major problem or failure.

At the lower voltages the application of periodic steep pulses [10] in, for example, HV motor-control devices may require special insulation monitoring systems. Also, sensitive local distribution networks could justify the development of new techniques.

Descriptions and analyses of some of the more advanced industrial monitoring methods now in use or under development are described in later chapters. These follow a review of the materials applied, their location in particular power equipment, possible fault conditions and details of established insulation assessment techniques.

1.6 Summary

In this introductory chapter the principle of an AC high-voltage power system is presented together with an indication of the types of equipment involved. The order of magnitude of the operating and associated test voltages are reviewed. The concept of insulation coordination for protecting power equipment insulation from damage due to lightning and switching surges is described. In addition, the developments in HVDC transmission are considered, including recent progress in localized schemes. Appropriate references are given. The need for future insulation assessment and monitoring is emphasized.

1.7 References

1. Jenkins, N., 'Impact of dispersed generation on power systems', *CIGRE Electra* 199, December 2001, pp. 6–13
2. Weedy, B.M., and Cory, B.J., *Electric Power Systems* (John Wiley & Sons, New York, 1998)
3. Luckett, M., 'How VARs can travel', *IEE Review*, September 1999, pp. 207–10

Again, assessment of the condition of the insulating materials can be an extremely useful tool in acquiring data for the overall costing procedures. This is especially effective if methods for continuous monitoring from new are incorporated in critical plant items, thereby enabling 'trend' statistics to be obtained for the equipment type and insulation system.

1.4.3 Extension of power system life

As part of the procedures related to the development of a system it is highly desirable to consider techniques for extending the lives of equipment beyond their expected retirement date. Much work is proceeding with the reconditioning of plant, including large generators and transformers. The restringing of overhead lines and the replacement of items such as cables, instrument transformers and switchgear become necessary for life extension. The appropriate time for commencing such work depends on many factors, the state of the insulation being of major significance. Reliable and informative data for insulation condition monitoring are essential as part of the decision-making process. A range of established and newer techniques are described in Chapters 6–10.

1.4.4 New systems and equipment

In the immediate future, power system technological developments involving insulation will include more extensive use of composite insulators for the overhead lines, the increased application of gas-filled equipment possibly with minimization of the SF_6 content, utilization of fibre-optic instrument transformers and installation of plastic – in particular XLPE – cables up to the highest voltages.

Some of the newer developments could include the application of 'cable' wound generators, motors and transformers for direct connection to the transmission/distribution systems [15] and the use of high-temperature superconductivity (HTS) cables [16] for large power transfers in metropolitan areas. HTS research and development are well advanced in the USA, Europe and Japan.

Considerable activity is continuing in the development of HVDC as a major contributor to the overall expansion in developing areas and as a backup to existing networks. On a smaller scale, the technology is being applied for local interconnections. A particular example of a small-scale development ($\pm$ 80 kV, 50 MW, 70 km XLPE cables) based on voltage source converter (VSC) technology is described in Reference 7. The 220 MW VSC Murraylink interconnection between Victoria and South Australia incorporates a 176 km land cable and the 330 MW VSC scheme from the USA mainland to Long Island a 40 km cable [17]. A comprehensive review of the components comprising a VSC scheme and possible operating characteristics are described in the CIGRE Brochure No. 269 [18] prepared by members of WG B4.37. Although experience with VSC systems is limited, the report is optimistic about the future of the technology. The highest ratings of systems in service (2004) were $\pm$ 150 kV and 350 MVA.

The various developments will inevitably require new insulation assessment and monitoring techniques.

the charging currents are too high for normal AC test equipment. HV DC test sources are also incorporated in HV impulse generators. At lower voltages DC techniques are applied in the measurements of a number of material dielectric properties as required for monitoring the insulation condition.

1.4 Power system developments

During recent years, in order to achieve increased utilization of the existing power system infrastructure, there has been a trend towards operating plant and line components much nearer to their maximum ratings and for longer periods than previously considered advisable from an engineering aspect. Such developments tend thermally to stress the equipment more highly and probably result in the specification of fewer planned time-based routine maintenance outages. The success of the changes will depend greatly on maintaining the insulating materials and structures in good condition. This applies to the more highly stressed designs now being introduced as well as to the ageing equipment that is required to remain in service.

1.4.1 Reliability requirements

An acceptably reliable system operating at or near full load demands that existing and new components must be monitored effectively. The manner in which this is achieved in relation to the system development and its overall operating and replacement costs is subject to much planning, necessitating complex decisions by the asset managers of present-day power supply systems.

The reliability requirements expected in a system depend on the continuous availability of adequate generation and the efficient maintenance of the supply network incorporating overhead lines and cables, together with the various associated plant items.

Statutory regulations specify that the voltage levels must be maintained during changes in the load conditions. Failure to meet the contractual requirements can result in high economic penalties. This will apply if a shutdown is necessary and may include liability for environmental damage due to malfunctioning of plant. As far as possible the safety of personnel must be ensured.

Some applications of artificial intelligence for incipient fault diagnosis and condition assessment are discussed in Chapter 10. It is clear that effective methods for monitoring of the various insulating materials must be included in the economic and technical development of existing and new power systems.

1.4.2 Condition of present assets

In many organizations system development includes the determination of the value of the existing assets in terms of predicted remaining life. This is a difficult process with plant ranging from ages of more than 25 years, perhaps on the upturn of the 'bathtub' curve, through to modern equipment of high capital cost incorporating updated designs and some new types of materials.

3. Discuss the reasons for the various high-voltage tests as applied to power-system equipment. Indicate why and on what bases were the relative levels and forms established by the industry.
4. Describe the advantages and disadvantages of HVDC transmission systems, including their relationship to the AC systems. Discuss the application of the newer developments involving VSC technology.

Chapter 2

Insulating materials utilized in power-system equipment

- The main insulating materials
- Characterization of insulation condition
- Modes of insulation deterioration and failure
- Electrical operating stresses

The successful operation of high-voltage power-system equipment is very dependent on the correct choice of insulating materials and maintaining them in good condition throughout their life. This requires knowledge of the types of traditional and contemporary materials available and how they would be expected to behave in the particular operating environment, especially over long periods.

The acceptance by the industry of new materials is a slow process because of cost restrictions, changes in production techniques and the requirement that a high probability of reliable continuous performance for periods exceeding 25 years can be achieved. Such estimates are based on experience, experimental test results and statistical analyses. Numerous test specifications have been written in order to assist in determining such reliabilities. Some of the methods are discussed in Chapter 6. The most important developments with the newer materials have been associated with the use of SF_6 gas in switchgear and transformers, plastics in high-voltage cables and different forms of synthetic polymers and glass fibres in machines, power/instrument transformers and insulators.

In this chapter a range of insulating materials and their special areas of application in the power system are reviewed. This includes the well-established materials, as these still form much of the insulation, i.e. air, hydrogen, wood, porcelain, glass, hydrocarbon oil, oil-impregnated paper, oil-impregnated pressboard, wrappings of synthetic-resin-bonded paper, resin-bonded wood laminates, resin-bonded paper laminates, and the newer materials.

In the review of insulating materials (Section 2.1) reference is made to the electrical parameters – permittivity (ε), resistivity (ρ), dielectric dissipation factor

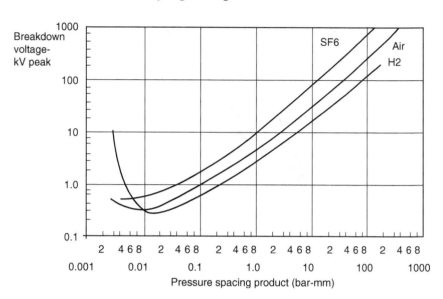

Figure 2.2   Approximate breakdown strengths of gases in a uniform field – 'Paschen Curves'

gases exhibit increased electric breakdown stresses at small spacings and in near uniform fields [7]. The approximate relationships based on data published in *Electra* Nos. 32 and 52 are depicted in Figure 2.2, which also includes results for hydrogen. In these types of fields once a partial discharge event occurs complete breakdown would be expected. This is especially significant with direct voltages, as, for example, within the air gaps of impulse generators (see Chapter 6).

The relationship for air [8] can be approximately represented by the equation

$$V_{peak}(kV) \approx 6.72\sqrt{pd} + 24.36pd \qquad (2.1)$$

where (pd) is in bar-cm. The equation is valid for pd >0.1 bar-mm.

The reduction in strength for non-uniform field conditions for air is indicated in Figure 2.1. In such configurations high stresses are produced at sharp electrodes and will trigger the ultimate flashover. However, partial breakdown can be maintained without failure in some gap types resulting in the phenomenon of corona. This may be observed visually in air, for example on overhead lines and insulators, and has been studied very extensively from basic physical characteristics [8] through to the effect of the associated finite losses on the performance of the lines.

Nitrogen is used at pressures up to 1.0 MPa in standard capacitors and in some forms of cables, while the low density of hydrogen is exploited in large water-cooled turbo-generators. As shown in Figure 2.2, the breakdown strength of hydrogen at atmospheric pressure is about half that of air. Operating pressures are in the region of 0.4/0.5 MPa with moisture contents corresponding to dew points of the order of $-20°C$.

2.1.2   Vacuum

In its pure state a high vacuum is an ideal dielectric over short distances, since no electron multiplication is possible. However, in practical equipment such as high-voltage circuit breakers, contamination from the metallic and insulation surfaces, together with residual oil and gases, limits the voltage stresses that can be achieved [9]. With good design and appropriate electrode materials, vacuum circuit breakers are now used in circuits up to and including 36 kV.

2.1.3   Liquids

The use of oil as an insulant is very common, either on its own or as an impregnant for achieving the good properties of a laminated or porous 'solid' material – in the case of transformers and some designs of high-voltage cables it acts as a heat-transfer medium between the active conductors and water or air coolers.

The type and quality of oil required is dependent on the particular application. The specifications range from normal hydrocarbon oils as supplied for use in switchgear and transformers through to special types for cables and capacitors [10]. When used as impregnants, the liquids are carefully dried, degassed and filtered to produce structures of high dielectric strength.

The motivation for using 'paraffinic'-based oils in place of 'naphthenic' types seems to have diminished. Much work was carried out on the former, especially with respect to viscosity at low temperatures and ageing characteristics [11].

The qualities of the various oils are checked by tests laid down in specifications [e.g. S2/5–S2/7] but the results do not necessarily indicate how the liquids will behave when built into a complex structure over a long period.

A simple example is the reduction in partial discharge and breakdown stresses with increase in spacing or volume in a uniform field gap as indicated in Figure 2.3 for 50 Hz and impulse voltage conditions. The graphs are representative of a wide spread of results from several sources, as reviewed in reference [12], and are included here only as an indication of an important trend when applying data from a particular sample to a configuration of different dimensions. The absolute stress magnitudes will depend on the condition of the oil tested or being used in the equipment.

During recent years concern has arisen regarding the effect of particles on the strength of oil as utilized in high-voltage power transformers. Some earlier experimental results [13] showed that an increase in the density of suspended particles ($>5\,\mu$m) from $2\,000/100\,cm^3$ to $12\,000/100\,cm^3$ reduced the breakdown voltage of large oil volumes by the order of 40 per cent. The test electrodes were concentric cylinders and the oil volume approximately $4 \times 10^5\,cm^3$. The average breakdown stresses for low-particle contents were of the same order as indicated in Figure 2.3 (b). A later report by CIGRE [14] contains data from 15 laboratories and an indication of the particle concentrations to be expected in practical transformers. Such data are applied in design, although further information is probably required relating to partial-discharge inception stresses, especially under surge voltages.

The macroscopic behaviour of oil in small and long gaps has been extensively researched and is still the object of experimental studies, e.g. References 15 and 16.

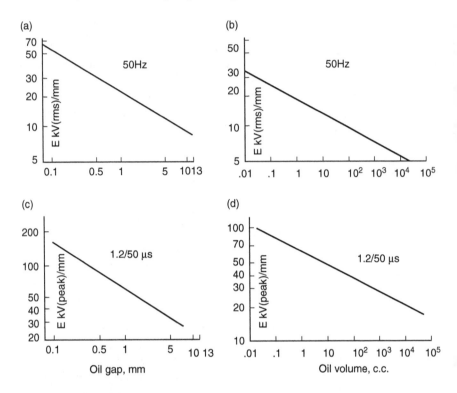

Figure 2.3   Reduction in oil partial-discharge inception and breakdown stresses in uniform fields – range of experimental results [12]. (a) 50 Hz PD inception stresses; (b)–(d) Breakdown stresses

This work is aimed at understanding the breakdown mechanisms and, more importantly, the partial-discharge phenomena with and without contaminants – in particular moisture, conducting and non-conducting particles and air.

As the gas-absorbing characteristics of oils can vary, some difficulties may arise in the interpretation of gas-in-oil analyses. These gas-absorbing oils are now used in transformers as well as cables and capacitors. In the latter cases they help minimize the formation of bubbles in the tightly packed insulation structures. A new test incorporating a point-sphere electrode system for checking the partial discharge characteristics of oils has been developed by CIGRE (SC 15) and is now an IEC Standard [S2/8].

An oil characteristic that continues to be of interest is concerned with the electrostatic charging effect produced by flow rates of, perhaps, 1.5 metres/second within the insulation configuration of certain designs of high-voltage power transformers. Extensive studies have been made in Japan, the USA and Europe. It appears that localized charges can be built up on insulation surfaces that may produce dangerous partial discharges and even flashover at the interface or in the bulk oil. Techniques have been developed for assessing the electrostatic charging tendencies (ECT) of the different oils (new and aged) on their own [17] or in conjunction with pressboard surfaces [18], as also investigated by Study Committees 12/15 of CIGRE [19].

Synthetic liquids were introduced for use in distribution transformers many years ago in order to overcome the fire hazards associated with hydrocarbon oils. One of the commoner liquids consisted predominantly of polychlorobiphenyl (PCB), a substance that is now unacceptable for environmental and health reasons. Extremely low concentrations are allowed in existing equipment – for example <0.5 ppm.

The destruction of PCB liquids has required the development of special collection and waste-disposal techniques, at considerable cost to the electrical industry. An advantage of PCB liquids was their high permittivity when used in capacitors. Replacement oils, including silicone liquids, are now available for use in smaller transformers and, for power capacitors, a wide range of individually designed synthetic oils has been developed [15, 16]. A well-established synthetic liquid is dodecylbenzene (DDB) for impregnation of wrapped insulation in high-voltage cables. It is claimed that the liquid has better ageing and gas-absorption characteristics than the natural oils. If low-temperature cables prove to be commercially viable, it appears that liquid nitrogen and/or helium will be considered as possible fluids for impregnation of lapped plastic dielectrics.

2.1.4   Solids

Clearly the structures of all the power-system equipment must include solid insulating materials capable of efficiently supporting, and isolating from each other, conductors at different potentials. Such materials must have adequate puncture and creep/tracking strengths. They must be able to withstand the expected thermal, mechanical and chemical conditions and maintain electrical stresses to give economic and technically acceptable designs. The orders of magnitude of achievable stresses are summarized in Section 2.4 for sample and equipment configurations. The materials reviewed include a limited number of the more important synthetic polymers applied in power system engineering. The industrial development of these materials is continuous but the insulation engineer must be cautious when assessing a new material for a particular application. The well-established might be the best solution, especially in relation to long-service performance.

2.1.4.1   Wood

Wood is one of the oldest insulations used by electrical engineers and despite limitations in the natural form it is widely applied. The outstanding application is in overhead line systems, where its relative cheapness and insulation properties are attractive [20]. It is also utilized in transformers, some older switchgear and generators in a laminated form suitably dried and glued/impregnated with resins to give a high mechanical strength and acceptable electrical properties.

2.1.4.2   Porcelain

For many years porcelain, and to a lesser degree glass, had no competitor as an insulation for overhead line insulators. It weathers well, even under moderate pollution,

for leakage currents, thus minimizing the possibility of dry-band arcing. Silicone is probably the only polymer having this very desirable property.

Liquid silicone rubber (LSR) appears to be advantageous where complex shapes are required and avoidance of voids is essential [23]. Room-temperature-vulcanized silicone rubber (RTV) has been applied to porcelain insulators in order to improve their wet and pollution performance [24]. A comparison of the properties of EPDM and HTV silicone rubber is given in Reference 1. Dielectric constant values are of the order of 2.5–3.5 and 3.3–4.0 respectively.

2.1.4.8 Heat-shrinkable materials

An important development was the introduction of heat-shrinkable polymeric materials [25]. This led to changes in the techniques adopted for 11 kV (and above) cable terminations at switchgear and similar locations. Much testing has been completed in the laboratory and at outdoor test sites. Assessment included the determination of the behaviour of the shrunken material when subjected to thermal cycling, as in a cable. Air gaps must not appear between the sleeve and the cable insulation (plastic or oil-impregnated), as this could result in partial discharges with subsequent failure (see Chapter 4).

2.1.4.9 Polyethylene (including XLPE)

The use of polyethylene as an insulating material is attractive, because it has low losses and moderately high electric strength. Initially, its thermal stability was unacceptable at the temperatures required in power engineering and not until cross-linked polyethylene (XLPE) was introduced did it find wide application [1]. The material can be extruded and was found suitable for cable manufacture once a number of problems were solved. These included development of methods for curing and cooling long lengths, at the same time eliminating voids in which PDs might be initiated. XLPE is susceptible to discharges of the order of tens of pCs but is now being applied as the major insulation in cables over a wide range of voltage, a few as high as 500 kV (see Chapter 4). The dielectric constant is 2.3.

2.1.4.10 Polyvinyl chloride

Polyvinyl chloride (PVC) is used extensively for the insulation of wires and as a sleeve material for low-voltage applications. In many power situations the integrity of the secondary wiring is vital. PVC has a thermal rating of 105°C, which can be increased by suitable formulations, and a dielectric constant of 3.0–4.0, depending on the form. The material is resistant to a range of liquids but is attacked by others, in particular aromatic hydrocarbons possibly associated with migration of plasticizer from the PVC [1].

2.1.4.11 Polytetrafluoroethylene

Polytetrafluoroethylene (PTFE) extrusions, moulds and films are used where demanding dielectric, mechanical, chemical and thermal conditions are encountered. Specialist applications include insulators, cables, wires and windings where the high cost can be justified. The material is susceptible to corona and radiation. Its dielectric constant is 2.0. Fluorinated ethylene propylene copolymer (PEP) – Teflon – has similar properties to PTFE but is not as tough and has a more limited temperature range. It may be processed by conventional extrusion and moulding methods.

2.1.4.12 Polypropylene

Polypropylene (PE) film is one of the materials used as the dielectric in power capacitors, having high electric strength and low losses. It has a low dielectric constant and the appropriate grade must be selected to minimize swelling in some dielectric liquids as used in capacitors.

In its bulk form the material is utilized for moulding of components and extruded as insulation for cables operating at less than 5 kV [1]. Polypropylene tapes laminated with paper (PPL) have been used for high-voltage cable insulation for ten years or more – for example by the National Grid in the UK.

2.1.4.13 Liquid-impregnated insulation systems

At the higher voltages it seems that liquid-impregnated systems will continue to be used successfully at extra-high voltages in cables, bushings, instrument transformers, power transformers, distribution transformers and power capacitors, although gas systems (SF_6) are now well developed and are competitive in a number of applications. Despite its apparent disadvantages, oil-impregnated paper (OIP) has proved a reliable and economic insulant in many applications. The achievement of the efficient utilization of natural materials has resulted from R&D effort over many years – in particular, the determination of conditions necessary to withstand high electric stresses for the expected lifetimes of at least 20 to 30 years. Of special interest to the design and operating engineers are moisture content, gas content of the impregnant, losses at operating stresses, partial-discharge inception stresses, the location of any PDs and the ageing characteristics.

Initial drying and impregnation conditions necessary to avoid failure are well established and form part of the production know-how. Vacuum ovens capable of pressures down to 0.1 torr or lower and temperatures of 100°–130° are common. The processing times of 2–3 weeks for a large transformer may be shortened very considerably by installation of vapour phase heating equipment, for example by using kerosene for heat transfer in the early stages of dry out. It is common for dryness levels of less than 0.1% to be achieved in such processes before impregnation. The oil would be expected to have a moisture content of a few ppm (e.g. <2 ppm).

The losses may be related to moisture content as well as to the quality of the material: a low-loss paper is used in cable manufacture. An effect found when attempting to establish breakdown criteria for new OIP is indicated in Figure 2.4. It is essential that low moisture contents be maintained in the practical systems if thermal runaway conditions and PDs are to be avoided (see Section 2.3). This is particularly important with cables in which the conductor losses pass through the OIP and add to the dielectric losses.

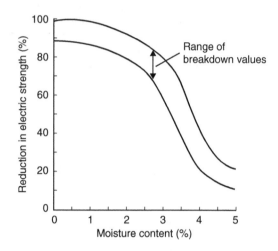

Figure 2.4   Effect of moisture on electric strength of oil-impregnated paper insulation

In high-voltage cables and some current transformers, arrangements are made to seal the system against the atmosphere, maintaining a moisture level of 0.5% or less, while in power and distribution transformers the situation is less critical, the stresses being lower and the insulation structure not so compact. This also means that gas-absorbent oils are not usually required in transformers, although there are indications that such oils are being provided as the norm. In cables, power capacitors, instrument transformers and certain regions of power transformers the existence of a gas bubble can be very significant, as it may lead to PDs that are confined, eventually leading to deterioration of the tape or sheet material. In the degradation process it appears that low-level ionization can produce a form of cross-linking in the oil impregnant resulting in the formation of the well-known X-wax compound containing carbon and hydrogen. After disassembly the presence of the compound is often taken as an indication that discharging has occurred in the paper/oil system. X-wax may be detected by magenta dye or ultraviolet light. It can be observed as a smooth cheeselike substance within the surface of the paper insulation. The effect is discussed in References 26 and 27.

A special situation in operating oil-filled power transformers is where the moisture content has risen to perhaps 2.5–4 per cent and during an overload the conductor temperatures are allowed to increase to 140° (IEC 60354) or 180°, as specified in the IEEE/ANSI standards for higher-temperature materials. Under such conditions it seems possible that gas bubbles may be released resulting in partial discharges. The effect of this will depend on their location and the electric stress values. As it was suspected that a number of transformer failures may have been due to bubble formation during overload conditions [28], the phenomenon has been investigated in some detail, including a joint project between Monash University and Electric Power Research Institute. It was demonstrated in a separate research programme that problems in the transformer interwinding insulation might arise at the higher temperatures [29].

Extensive developments in power capacitor insulation systems included the introduction of polypropylene film in conjunction with the more traditional paper, the latter acting as an impregnating interface. However, for a number of years, all-film capacitors, in which a 'hazy' or similar polypropylene film is used to enable impregnation to take place, have been in service. Capacitor insulation is very susceptible to impurities because of the thickness of only 10 micrometres (μm) and the very high stresses (tens of V/μm) used. The dielectric losses in the new designs are very low and no longer a major limitation in design, giving low DDF values. The development of insulating materials for high-voltage capacitors continues to be active.

2.2   Characterization of insulation condition

A number of measurable parameters can be employed – directly or indirectly – to characterize the condition of insulating materials when built into system equipment. The predominant electrical characteristics are the values of permittivity and capacitance, resistivity and insulation resistance, insulation time constants, dielectric dissipation factor and partial discharge status, all of which are defined in Subsections 2.2.1–5. In oil-insulated apparatus the levels of moisture content, dissolved gas volumes and various chemical quantities are indicative of the oil/solid condition. These are discussed in Subsection 2.2.6 and in later chapters.

2.2.1   Permittivity (ε) and capacitance (C)

The relative permittivity, ε_r, is the ratio between capacitances of identical electrode systems with and without a dielectric present. The relationship is the capacitance $C = k\,\varepsilon_r\varepsilon_0$, where k is a constant representing the geometrical structure of the system and $\varepsilon_0 = 8.854\ 10^{-12}$. The values of ε_r range from 1 for air through to 5.5 or so for porcelain. Capacitance values range from a few pF (10^{-12} farads) for cap-and-pin insulators through to μF (10^{-6} farads) for cables, generators and power capacitors. In addition to the requirements of measurement systems (see Chapter 6) relative permittivity magnitudes are important in the alternating voltage design of multiple dielectric structures (see Chapter 3). At power frequencies little variation occurs in the value of ε_r with applied stress or temperature for the materials used in power system practice. The presence of moisture ($\varepsilon_r \approx 80$) or trapped gas might be expected to increase or decrease respectively the measured values.

2.2.2   Resistivity (ρ) and insulation resistance (IR)

Volume resistivity, ρ, is defined as the resistance between opposite faces of a 1-metre cube of the insulation.

The resistance $R = \rho\,L/A$, where L is the spacing between electrodes and A is the area of the electrodes. The unit of volume resistivity is the ohm-metre, the degradation. This is especially the case for oil-immersed systems, where detection of combustible gases and of furans in the oil has resulted in the development of specialized techniques for assessing the condition of the materials. The methods utilize DGA and the application of liquid chromatography (HLPC).

An important practical insulation condition indicator associated with oil-immersed equipment is the analysis of the gases and their relative concentrations as produced by particular types of fault in the structure. Possible diagnoses are quantified in the revised IEC Publication 60599 [S2/10] and discussed by Duval (see Chapter 6). The analyses give signatures indicating the presence of low-energy partial discharges through to high-energy faults due to arcing. Interpretation of the data obtained by such techniques is considered in Chapters 9 and 10.

The physical ageing of paper as represented by the reduction in the degree of polymerization (DP) has been related to the amount of furanic compounds produced at hot-spot temperatures and absorbed in the oil of transformers in service [30]. A reduction in the magnitude of the DP from a value of $>1\,000$ (2-Furaldehyde 0–0.1 ppm) when new to 250 (2-Furaldehyde >10 ppm) is considered to be the end of life for oil-impregnated cellulose by some authorities. (See also Reference 39 of Chapter 6.) Much work is being carried out in this field, including studies for assessing the characteristics of naturally aged oil-impregnated materials aimed at estimating probable life times – for example, Reference 31. The formation of water under certain conditions, such as in sealed units, may be of considerable importance in the prediction of the rate of ageing in oil – paper systems.

Conventionally, oil acidity, interfacial tension and sludging tendency are used as physical and chemical guides for monitoring the condition of transformer oil [S2/6].

Determination of the physical changes associated with the various forms of treeing and tracking in XLPE and resins has been the aim of much research. The understanding of the mechanism of water treeing in XLPE is of considerable practical significance and remains an important research topic. This work requires the removal of samples from equipment suspected of deterioration and no longer suitable for normal service and therefore is only an indirect monitoring technique.

2.3 Modes of deterioration and failure of practical insulating materials

The concept of insulating materials having an intrinsic breakdown strength was postulated many years ago. It was found that, for thin specimens of certain pure dielectrics and by minimization of edge effects in the test configuration, the impulse breakdown stresses could be increased to perhaps ten times that achievable with normal materials. However, in practice, impurities and manufacturing variations prevent exploitation at a large scale of these purely electronic processes.

The major deteriorative and failure modes associated with power equipment insulating materials, which often comprise combinations of solid and gas/air or solid and liquid, are listed below. This is followed by consideration of some of the processes involved. In the case of the physical and chemical changes, the present treatment is limited to mention of those aspects considered helpful in understanding the condition monitoring requirements.

(a) Dielectric losses causing thermal instability, or runaway, in the bulk of the solid material.
(b) Partial discharges representative of

- partial breakdown in voids and gaps enclosed within the solid insulation producing local erosion of adjacent material;
- partial breakdown in the ambient medium (oil, gas or air) at an interface between an electrode to dielectric or a dielectric to dielectric, thereby initiating flashover or creep and, perhaps, localized puncture through an adjacent surface.

(c) Ageing due to thermal, electrical and mechanical stressing, including the effects in (a) and (b).
(d) Long-term chemical changes produced by incompatibility between materials, resulting in the creation of dangerous by-products. The changes are also related to (c).
(e) Deterioration of surface material by external pollutants, leading to reduction in tracking strength.

2.3.1 Dielectric losses

The failure of insulating materials due to high dielectric losses is a well-known phenomenon. In practice the effect is due to the use of, for example, unsuitable resins and often to the presence of moisture. The latter may be produced by chemical ageing, in some cases because of poor equipment sealing and, in others, by incorrect processing procedures in the factory. Additionally, the summation of the effects of energy dissipated by any high-value partial discharges can contribute significantly to the overall dielectric losses.

2.3.1.1 Determination of loss relationship

A measure of the total losses may be obtained by considering the relationship between the applied voltage and the current flowing in a simple capacitor. The relevant phasor diagram is given in Figure 2.5 as defined in Subsection 2.2.4. Assuming an angular frequency of ω the loss under alternating voltage conditions is

$$W = VI_R$$
$$= VI_C \tan \delta$$
$$= V^2 \omega C \tan \delta \text{ watts} \qquad (2.2)$$

where V is the voltage in volts (RMS) and C is the capacitance in farads.

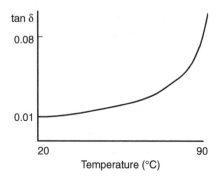

Figure 2.6   Possible variation of tan δ with temperature for multi-layer clamping board

If C is assumed to be a cube of metre length sides then $C = \varepsilon_r\varepsilon_0$ farads and equation 2.2 becomes

$$W = V^2\omega\varepsilon_r\varepsilon_0 \tan \delta \text{ watts/cu. metre}$$
$$= E^2\omega\varepsilon_r\varepsilon_0 \tan \delta \quad \text{watts/cu. metre} \tag{2.3}$$

where E = volts/metre, $\varepsilon_0 = 8.854 \times 10^{-12}$ and ε_r = relative permittivity.

In a stable direct-voltage system the losses would be simply represented by the I^2R values. Loss effects can be determined from calculations similar to the following.

2.3.1.2   Example of loss calculation

The relationship given in (2.2) may be applied for determining possible thermal instability in equipment. In studies related to the operation of oil-impregnated-paper power cables it was shown many years ago that the conductor losses, added to the dielectric losses, could lead to a continuous increase in the paper temperature. The calculation assumes a reference temperature on the conductor, e.g. 90°, an imposed outer screen temperature, knowledge of the variation of tan δ with temperature and the dielectric thermal conductivity. An allowance is made for the logarithmic variation of stress (E) through the insulation [26].

A simpler calculation has been made for an oil-filled transformer with wooden laminated winding clamping boards. In this case it was required to determine the order of magnitude of electric stress that would be expected to produce thermal instability in 5 cm- and 10 cm-thick boards, with the oil at a temperature of 90°. The variation of tan δ with temperature was assumed to be of the form in Figure 2.6. The calculated temperature distributions through the boards for a range of electric stresses for the two cases are indicated in Figure 2.7. The temperature rise due to a known loss in the dielectric may be determined from (2.4).

$$\Delta\theta = (1/\sigma_T)(d/A)\Sigma\Delta W \tag{2.4}$$

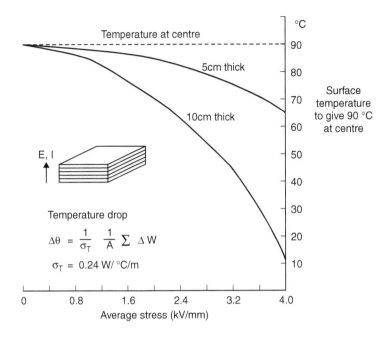

Figure 2.7   Calculated temperature distributions through insulation boards for variation in electric stress

where $\Delta\theta$ = temperature drop through a dielectric with thickness of d metre and cross-sectional area of A square metres. A thermal conductivity of σ_T = 0.24 watts/°/metre is assumed.

As an example, the dielectric may be divided into ten sections for the calculation and the magnitude of $\Delta\theta$ determined for each using the value of internal loss dissipated and that due to thermal flow through the section.

2.3.2   Partial discharges – sources, forms and effects

In order to appreciate the significance of partial discharges and how they can influence the life of the insulating materials it is necessary to understand where and how they might occur and the manner in which they can lead to deterioration and ultimate failure. These factors are considered briefly in the following.

2.3.2.1   Sources of partial discharges

Under alternating voltage conditions the PD sources are usually associated with trapped gas or the ambient medium. Such components of a combined insulation system have lower permittivities and breakdown strengths than the solid materials and therefore tend to be the initial location of the local partial discharges. This aspect is considered in Chapter 3 together with descriptions of some of the insulation configurations in various equipment. Despite the differences in the structures it is

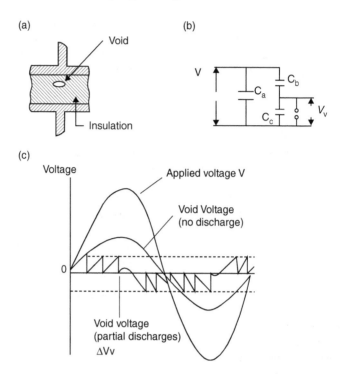

Figure 2.9   Simplified model of sample containing partial discharges [32, 33]. (a) Physical model; (b) Equivalent capacitance network; (c) Sequence of PDs for inception well below alternating voltage peak

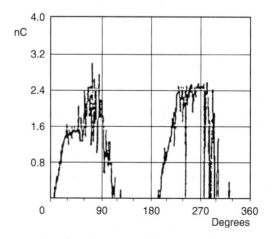

Figure 2.10   Typical PD pattern from a 'wet' pressboard sample

In Figure 2.9 the magnitude of the terminal pulse, ΔV_x, depends on the relative values of the capacitances C_c and C_b, which are usually unknown, and that of C_a. The apparent charge transfer (partial discharge) is normally the parameter measured at the terminals. The relationship between the cavity charge transfer and that at the terminal may be determined.

Referring to Figure 2.9b the change of charge across the void

$$\Delta q_c = \Delta V_c \cdot \{C_c + C_b C_a/(C_a + C_b)\}$$
$$\approx \Delta V_c\{C_c + C_b\} \text{ as } C_a \gg C_b \text{ (volume of } C_b \text{ is small)}$$
$$\approx \Delta V_c C_c, \text{ as } C_c \gg C_b \text{(area of } C_b \text{ is small and is length of sample)}$$

The terminal voltage pulse due to the discharge in the cavity is

$$\Delta V_x = \Delta V_c C_b/(C_a + C_b)(\text{capacitive division})$$

The apparent discharge Δq_x detected at the terminals is

$$\Delta q_x = \Delta V_x C_x = \Delta V_x\{C_a + C_b C_c/(C_b + C_c)\}$$
$$\approx \Delta V_x(C_a + C_b) \text{ as } C_c \gg C_b$$
$$\approx \Delta V_c C_b$$

Thus $\Delta q_x \approx \Delta q_c C_b/C_c$ \hfill (2.5)

which is the quantity normally measured at the terminals. It will be noted that it would be expected to be much less than that discharged in the void as, in general, $C_c \gg C_b$. However, for a given system and approximately known location the value of Δq_x is a good measure of potential damage. It is often described as the *apparent partial discharge*.

The partial discharges that can occur within the ambient near insulation surfaces – as in regions A of Figure 2.8 (iv) – are often of larger values than disturbances within the bulk material. In air-insulated systems it may be possible to calculate the order of magnitude of the inception stress containing a wedge adjacent to the surface [37]. For a sharp-edged electrode on a surface in oil or air (Figure 2.8(v, 1) and (v, 3)), Kind [36] quotes the empirical relationship

$$V_i = K(s/\varepsilon_r)^{0.45} \hfill (2.6)$$

where V_i is in kV, s is the spacing in cm, and the value of K is approximately as follows:

For 'corona' discharge inception $K = 8$ for metal edge in air

$K = 12$ for graphite edge in air

$K = 30$ for metal or graphite edge in oil

For 'brush' discharge inception $K = 80$ for metal or graphite edge in air or oil

2.3.2.3   Other characteristics

Because of the unknown differences in the relative charge values and the physical changes occurring, it is considered by many investigators that energy values have a greater significance. It can easily be shown (for example, in Reference 34) that for the idealized capacitive network and a single pulse the energy dissipated in the cavity is approximately equal to the measured terminal value of 0.7 $\Delta q_x V_i$, where V_i is the RMS value of the terminal inception voltage. In order to obtain a better estimate of the energy being dissipated, it is necessary to make a cumulative measurement of $\Delta q_x v_i$, where Δq_x is the terminal partial discharge value and v_i is the instantaneous value of the applied high voltage. Such measurements are possible using specially designed instruments or a general computer-based measurement system with appropriate software analysis programs.

These newer techniques allow more accurate determination of such significant PD characteristics as time of occurrence within the power-frequency cycle, repetition rate, polarity, cumulative energy dissipation and various statistical analyses of the data. The measurements are an extension of many earlier studies using less sophisticated equipment aimed at recognizing partial discharges under particular conditions. Modern digitized systems have also enabled the detection and analysis of the individual pulse wave shapes, which may have rise times of a few ns and a duration of tens of ns. In some cases with the presence of an insulation surface these times can be of the order of a microsecond. Research in such areas is now being pursued in many laboratories worldwide. Reference should be made to published papers, for example the publications of the DEIS of the IEEE and those of the International Conference on High Voltage Engineering (ISH) conferences.

In the traditional measurement systems (see Chapter 6), which are still widely applied in industry, and assuming the inception values recorded are near the peak of the high-voltage wave, energy levels tend to be comparable. This is especially so if the results are for tests on similar plant.

2.3.2.4   Damage due to partial discharges

It is apparent that certain combinations of stress and contamination are required for the onset of partial discharges under operating conditions. During overvoltage tests PDs may be initiated by incorrect design and/or manufacturing errors. Any damaging effects will depend on the materials involved and, in many cases, whether or not the by-products, especially gases, are contained locally in the structure.

In solid materials such as resins and polyethylene a form of internal tracking or treeing occurs, the stress at the tips of the trees tending to determine the progress of the branches. This effect is well documented with dangerous levels being as low as tens of pCs in some cases. With micaceous-based resin systems internal PDs are probably associated with the organic material but treeing is restricted by the mica flakes, which are able to withstand thousands of pCs for many years at operating stresses.

In laminated materials – resin- or oil-impregnated – containing gas-filled voids tracking tends to develop along the sheet, or tape, material and may exist for long periods of time if the direction of the primary component of stress is across the laminations of the built-up structure (Figure 2.8 (ii)). Experience indicates that erosion can develop without failure with hundreds of pCs present when the stress is normal to the laminations or barriers. However, failure would be expected as the effect of the trapped gases increases and high values of several thousand pCs develop at operating stresses. Numerous 132 kV synthetic-resin-bonded paper (SRBP) bushings have survived for many years under such conditions. In some oil-impregnated systems moderate values of PDs of, perhaps, 1 000–2 000 pC can be withstood for many months.

A situation of practical importance is where discharges are initiated on a creep surface due to the stress through the material as indicated in Figure 2.8 (iv) and (v). In such cases flashover may take place along the surface as a track develops. PD values can vary from tens of pCs to thousands of pCs on oil-immersed pressboard and on polluted air surfaces in outdoor substations.

With very intense fields as in Figure 2.8 (v, 2) tens of thousands of pCs can be withstood for long periods at an oil-impregnated pressboard surface with average stresses as high as 7–8 kV/mm. The concentration of energy produces pitting and degradation of the fibres [38]. The energy required to produce a given volume of gas from oil-impregnated pressboard was estimated in the experiments by Fallou *et al.* [39].

The condition shown in Figure 2.8(vi) is unlikely to exist under oil unless some extraneous metal object, frayed lead, perhaps a gross error in design or a manufacturing fault is present. PD levels of 100 pC or so may be detected at near inception rising to several thousands at higher stresses, depending on the configuration. If remote from ground, such disturbances can continue without failure, because the oil is self-sealing. This is also the case for air where the initial values may be as low as 10–20 pC but can be many thousands of pCs at higher stresses – as in air-insulated equipment of the power system. The difference in PD polarity between oil and air at inception is notable and sometimes useful in interpretation. Air corona in an enclosed space – such as a high-voltage machine – can have a deteriorative effect due to the creation of ozone.

Although an unusual source of partial discharging, the consequence of bad connections can be very significant. Figure 2.8 (vii) indicates the effect of a broken connection where discharging occurs across the break due to capacitance coupling to the high-voltage electrode. The magnitude of the PDs will be partly dependent on such coupling and may be of the order of thousands of pCs under adverse physical conditions.

Much of the earlier work related to identification of PDs and associated possible damage was developed following the introduction of commercial discharge detectors in the 1950s. Many of the results were collated and presented in the report by CIGRE [40].

The manner in which the discharge sources might be identified using modern measurement techniques is an area of active research, some of which is detailed in CIGRE papers and in particular a brochure prepared by WG D1.11 [41].

2.4   Electrical breakdown and operating stresses

An indication has already been given in Figures 2.1–2.3 of the breakdown stresses in gases and transformer oil for simple electrode configurations. It is difficult to define or specify the electrical breakdown strength of individual materials, although this quantity is clearly of major interest for design purposes. The differences in strengths between materials are large with the values for similar samples of a particular type of dielectric varying for a specified test condition. This variation requires the introduction of probability concepts when choosing test and operational safety factors. Figure 1.9 is an example of how the mean breakdown strength of some materials might be expected to reduce with increase of the duration of the applied voltage.

In Table 2.1 are listed the order of magnitude of the 50 Hz operating stresses for some of the materials utilized in the various items of power-system equipment. Precise values cannot be quoted, as these are usually confidential, often depend on the local insulation configuration and vary according to the practice of the particular manufacturer. A number of typical sample breakdown stress values are also included in the table. The majority of such results are obtained from short-term tests on thin materials using the methods specified in a series of standards – for example IEC and ASTM. Usually these test configurations are not representative of the overall working conditions, although a particular parameter may be simulated. The tests comprise part of the quality-control process of the materials as manufactured by the supplier. The sample breakdown results are quoted in the specification for the particular material (or liquid) but must not be used as such in the design of full-scale equipment.

The electrical stress magnitudes at failure for new materials are influenced by many parameters, including the thickness of the material, the type of electrode system and the form and duration of the applied voltage. The relationship between the breakdown stress in a simple test sample and that in a full-scale structure of similar materials is difficult to determine. Also, the ratio of the power-frequency short-term equipment-test value and the much lower operating stress necessary to give an acceptable probability of life is not easily estimated. The various equipment manufacturers and users have established ratios through the relevant national and international standards committees. Ratios have also been selected to include associated lightning and switching impulse test levels. Some details of test procedures are discussed in Chapter 7.

It should be noted that the operating stresses are often 50 per cent, or even lower, of the values applied during the short-term power-frequency tests and much less than those in the impulse tests when these are required. The estimated breakdown stresses for a practical configuration installed as part of power-system equipment must be significantly higher than the test levels. This is necessary in order to maintain a safe margin during high-voltage routine and type tests. The value chosen is based on experimental tests with statistical analyses of the results if sufficient information is available, possibly supported by the use of advanced condition monitoring techniques. The choice of safety factors is influenced also by the experience and know-how of the manufacturer and user, especially in respect of the long-term operating stresses. An important factor is the need to appreciate that breakdown testing of full-scale partial structures may not necessarily allow for problems associated with the manufacture and assembly of the final equipment. For items manufactured in large quantities a lower safety margin may be acceptable.

Table 2.1   Examples of operating stresses and breakdown values of associated samples of insulating materials used in power-system equipment

Equipment	Type of insulation	Range of 50 Hz operating stresses kV(RMS)/mm	Selected sample 50 Hz short-term breakdown stresses (average)	
			kV(RMS)/mm	Reference
Rotating machines	Resin-mica tape	3–4.5		
Power transformers	OI paper	3–3.5	75 (ramp), 300 μm 37 (40 min), 1 mm	[45], [46]
	OI pressboard	3–4.5	40	IEC 60243 [47]
	Oil	3 (5 mm) <1(large)	12 (5 mm), 3.9 (15 cm)	Figure 2.3 [13]
	Oil-solid creep surface	0.4–2.0	7–8 (6 mm)	[46]
	Inter-phase air	For ex.300 kV Class		IEC 60076-3
		0.133 (2.250 m)	50cm. SG @300 kV 2.35 (180 mm)	IEC 60052
			Rod-plane @ 530 kV 0.265 (2.00 m)	Figure 2.1
Bushings	OI paper	3–4.5		
	SRBP	2		
	Resin-impregnated paper	3.5		
	Cast resin	3		
	Porcelain			
	– radial		4–16 (3 mm)	
	– surface (oil)	0.5		ASTM D149 [1]
	– surface (air)	0.2		
Current transformers	OI paper	3.5–4	As above	
	Cast resin	3		
Capacitors	Oil–paper	10–25		
	Film type	50		
Cables	OI paper (pressure)	12–15		
	– joints	7		
	XLPE	10–15 (new types) 4–8	48 (XLPE)	[48]
	EPR	3.5	20–32 (EPR)	ASTMD149 [1]
	Polypropylene paper-fluid (PPL)	10		

Continued

Table 2.1   Continued

Equipment	Type of insulation	Range of 50 Hz operating stresses kV(RMS)/mm	Selected sample 50 Hz short-term breakdown stresses (average)	
			kV(RMS)/mm	Reference
Insulators	Porcelain (creep)	20 mm/kV with medium pollution		
	Composite (creep)	0.5 (near termination)		
OH lines and air clearances	Air – small gaps		2.1 (10 mm UF gap)	Figure 2.2
	Air – long gaps	≪ 1	1.06 (1.50 m ≈ UF gap)	IEC 60052
	Air - non-uniform		0.28 (1.50 m gap)	Figure 2.2

2.5   Development of insulation applications

In conclusion of this chapter it is interesting to note that there appears to be an upsurge in activity related to the application of insulating materials in power-system equipment, especially in respect of the most efficient use of present plant and the incorporation of modern insulations in new equipment.

This ranges from attempting to estimate the remnant life of existing materials through to applications of the newer materials, e.g. XLPE high-voltage cables (up to 500 kVRMS or higher), SF_6 for insulation of highly rated transformers, low-loss and less hazardous liquids, metallized films for power capacitors and materials for use at superconductivity temperatures.

As discussed in later chapters, many techniques are available or under development for monitoring the critical insulating materials as installed in power-system equipment.

2.6   Summary

The review of insulating materials includes traditional and new forms as applied in power-system equipment. The electrical and physical properties of significance in characterizing and assessing the condition of the materials are introduced. A number of the deteriorative and failure mechanisms associated with practical insulating materials are described, including an indication of the magnitude of the electrical breakdown stresses of samples/prototypes compared with the actual operating stresses. An understanding of these various factors and the expected behaviour of the materials enables the most appropriate techniques for insulation-condition assessment to be selected and assists in interpretation of the complex data recorded by the monitoring systems. The latter are considered in later chapters.

2.7 References

1. Shugg, W. Tillar, *Handbook of Electrical and Electronic Insulating Materials*, 2nd edn (Van Nostrand Reinhold, NY, 1995)
2. *Engineering Dielectrics*, Vol. I Corona, Vol. IIA Solids, Vol. III Liquids' ASTM
3. Bradwell, A. (ed.), *Electrical Insulation* (Peter Peregrinus Ltd, London, 1983)
4. Ryan, H.M. (ed.), *High Voltage Engineering and Testing*, 2nd edn (IET, London, 2001)
5. *EPRI AC Transmission Line Reference Book - 200 kV and above*, 3rd edn (EPRI, Palo Alto, 2005)
6. Holmberg, M.E., and Gubanski, S.M., 'Motion of metallic particles in GIS', *IEEE Electrical Insulation Magazine*, July/August 1998: 5–14 (also September/October 2000)
7. Dakin, T.W., Luxa, G., Oppermann, G., Vigreux, J., Wind, G., and Winkelnkemper, H., 'Breakdown of gases in uniform fields: Paschen Curves for nitrogen, air and sulphur hexafluoride', *Electra*, January 1974;**32**:61–82 (see *Electra 52*, for hydrogen values and Heylen, A.E.D., *IEEE Electrical Insulation Magazine*, May/June 2006;**22**(3):25–35 for formulae)
8. Kuffel, E., Zaengl, W.S. and Kuffel, J., *High Voltage Engineering Fundamentals*, 2nd edn (Butterworth-Heinemann, UK, 2000)
9. Latham, R.V., *High-Voltage Vacuum Insulation: The Physical Basis* (Academic Press, London, 1981)
10. Wilson, A.C.M., *Insulating Liquids: their uses, manufacture and properties* (Peter Peregrinus, Stevenage, UK and NY, 1980)
11. Rouse, T.O., 'Mineral insulating oil in transformers' *IEEE Electrical Insulation Magazine*, May/June 1998: 6–16
12. James, R.E. 'Behaviour of oil immersed surfaces when subjected to tangential electrical stress at high voltage', PhD thesis, University of London, 1974
13. Samat, J., and Lacaze, D. 'Micro-particles in transformer oil and dielectric withstand effects', *Alsthom Review* 1988;(11):47–57
14. CIGRE Brochure 157, 'Effect of particles on transformer dielectric strength', *Electra*, June 2000;(190):135–40 (WG 12/17)
15. Beroual, A. *et al.*, 'Liquid Dielectrics Committee International Study Group', *IEEE Electrical Insulation Magazine*, March/April 1998: 6–17
16. 'Dielectric Liquids', special issue, *IEEE Transactions on Dielectrics and Electrical Insulation*, February 2002, vol. 9
17. Oommen, T.V., and Petrie, E.M., 'Electrostatic charging tendency of transformer oils', *IEEE Transactions on Power Apparatus and Systems*, 1984;**103**:1923–1931

Part 4-1 (Ed. 3.0, 1990): Part 4-2 (Ed. 1.0, 2000) and Part 4-3 (Ed. 1.0, 2000): Ageing ovens
Part 5 (Ed. 2.0, 2003): Determination of relative thermal endurance index (RTE) of an insulating material
Part 6 (Ed. 1.0, 2003): Determination of thermal endurance indices (TI and RTE) of an insulating material using the fixed time frame method

S2/3 IEC 60505 (Ed. 3.0, 2004): Evaluation and qualification of electrical insulation systems
S2/4 IEC/TS 61251 (Ed. 1.0, 1993): Electrical insulating materials – AC voltage endurance evaluation – Introduction
S2/5 IEC 60156 (Ed. 2.0, 1995): Insulating liquids – Determination of the breakdown voltage at power frequency – test method
S2/6 IEC 60296 (Ed. 3.0, 2003): Fluids for electrotechnical applications – unused mineral insulating oils for transformers and switchgear
S2/7 IEC 60247 (Ed. 3.0, 2004): Insulating liquids – measurement of relative permittivity, dielectric dissipation factor (tan δ) and DC resistivity
S2/8 IEC/TR 61249 (Ed. 1.0, 1993): Insulating liquids – Determination of the partial discharge inception voltage (PDIV) – Test procedure
S2/9 IEC 60270 (Ed. 3.0, 2000): High-voltage test techniques – partial discharge measurements
S2/10 IEC 60599 (Ed. 2.0, 1999): Mineral oil-impregnated electrical equipment in service – guide to the interpretation of dissolved and free gases analysis

2.9 Problems

1. How can moisture and/or air (gas) affect the electrical properties of extruded XLPE, oil-impregnated paper and resin mouldings as applied in high-voltage equipment? Outline the principal methods that might be used to determine the presence of such contaminants.
2. Determine the AC breakdown stresses in a simple uniform field insulation system with an overall spacing of 12 mm and an insulation resin covering of 2 mm thickness on one electrode when immersed in (a) air, (b) SF_6 (both at atmospheric pressure) and (c) dry transformer oil. Calculate stress in each 'ambient' medium and the resin. $\varepsilon_{resin} = 3.5$, $\varepsilon_{oil} = 2.2$
3. An air-filled cavity of 3 mm diameter and thickness of 1 mm is suspected of existing within the laminations of an insulating board of 40 mm thickness and large area. Estimate the magnitude of partial discharges that might exist in the cavity on application of a high voltage, V, across the board. What value of PD should be detectable at the electrodes? Determine the value of the applied voltage V required to initiate discharges in the cavity. $\varepsilon = 4.5$

Chapter 3

Introduction to electrical insulation design concepts

- Basic requirements for insulation design
- Electric stress distribution in insulation systems
- Electric stress control

During the process of choosing an appropriate insulation monitoring or assessment system it is of value to be aware of the design, manufacturing, electrical testing and operating requirements of the power equipment to be monitored. In this chapter an indication is given of some of the more important generic factors to be considered when selecting insulating materials for practical designs. The applications of suitable materials for a range of plant items are described in Chapters 4 and 5. The materials are reviewed in Chapter 2.

The interpretation of recorded data generated by potential faults and changes in the characteristics of the insulating materials may be aided by analysis of simple electric field configurations representing the original system. For complex arrangements it is often possible to simplify an area of interest to assist with a preliminary analysis. Several simplified cases are described in this chapter. Detailed analyses involve computerized field determinations based on accurate knowledge of the particular design. A special feature of electric field design is the use of stress-control techniques a number of which are listed together with reference to examples included in other chapters. The incorrect application of these methods can result in insulation problems detectable by appropriate monitoring systems.

3.1 Overview of insulation design requirements

The following presentation aims at giving an overview of the complexities involved in producing a satisfactory insulation structure capable of operating reliably for 25 to 40 years or more. Such lifetimes can be achieved only if efficient maintenance programmes, including insulation monitoring, are rigorously implemented.

3.1.1   Electrical requirements

The electrical test voltage/operating voltage ratio varies according to the system's basic insulation level, type of equipment and its location. The value is specified by the user, guided by the data in IEC 60071-1 (Chapter 1). Depending on the particular circumstances, partial discharge, radio interference voltage and dielectric loss levels must be below specified values as manufactured. To ensure these requirements are met, appropriate tests are applied as included on the diagram. Special arrangements may be required for in-service monitoring of, for example, PDs and DDF.

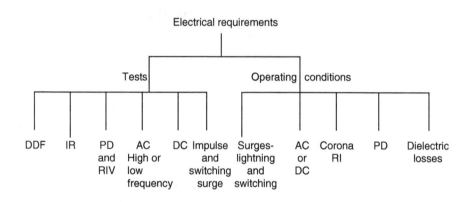

3.1.2   Physical limitations

Awareness of the physical limitations imposed by non-insulation factors is important. There are many occasions when such restrictions exist and only minor modifications are possible to improve the overall dielectric performance. Some of the physical changes to the initial overall design proposals that might be considered are indicated below.

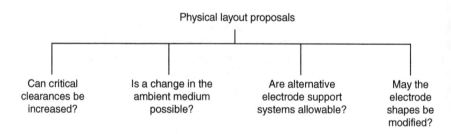

3.1.3   Working environment

No matter how well designed the structure and how satisfactory the test laboratory results, these measurements are of little value unless attempts have been made to simulate and/or allow for possible insulation deterioration due to poor working conditions. If the environment results in control equipment becoming covered with dust in a damp atmosphere, or a rotating machine contaminated with grease and dirt, or an overhead line insulator polluted with salt and industrial fumes, then these factors must be considered at the design stage.

A major deteriorative effect is due to the trapping of air or the build-up of gas leading to partial discharges and ultimate failure. This can occur in oil-insulated transformers due to poor impregnation, failure to 'top' up correctly, or not following specified procedures when installing a bushing. The ingress of moisture over a long period can increase the dielectric losses in many materials, again resulting in breakdown. This applies to solid materials such as epoxy resins and XLPE as well as oil-impregnated paper, pressboard and wood. Acceptance and monitoring measurements have been devised to cover the various conditions. These techniques are described in later chapters.

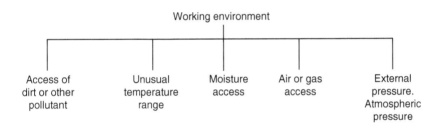

3.1.4   Mechanical requirements

As in much of electrical engineering, it is important to appreciate that allowance for the mechanical requirements in the application of insulation is essential. This is often the prime factor in ancillary low voltage systems including coaxial cables, wiring, plugs and sockets and some windings. At high voltages, in such items as rotating machines, suspension insulators, switchgear, transformers, compressed-gas or vacuum equipment and cables the mechanical performance of the insulation is of major consequence.

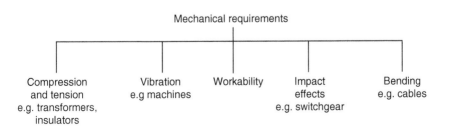

3.2   Electric stress distributions in simple insulation systems

There are many difficulties in applying the data available concerning the dielectric characteristics of insulation, particularly the results from small sample tests where the concepts involved in volume, area and thickness effects must be considered. Allowance needs to be made for any variations with time of electrode voltages, as can occur when a transformer winding is subjected to a lightning impulse or, sometimes, where direct and alternating voltages are applied simultaneously.

Because of the many conflicting factors it is often impossible to configure the materials to obtain an optimum insulating system. In such cases it is necessary to prepare a compromise arrangement within the overall equipment design and production limitations, as well as the customer specification. Considerable expertise is required in the layout of insulation to enable a proposed electrode system of high-voltage and earthed components to withstand the various test and operating conditions. To this end knowledge of the electric field distribution is essential together with information regarding the application of sample test results to the practical configuration, electric strength data and, perhaps most importantly, previous experience accumulated by the organization.

Design of insulation structures is often assisted by consideration of simple configurations, which can be helpful in simulating the region of interest. For some arrangements it is difficult to interpret a field determined by an advanced three-dimensional computer program unless prior knowledge as to the weakest region in relation to the local stress is known. The highest stress may occur in an inherently strong insulation structure but elsewhere a dangerous lower stress might exist in a location that is basically very difficult to insulate. Such features are more obvious in the well-established, but now rarely used, analogue techniques set up for the plotting of capacitive fields in conductive media, the commonest two being a resistive paper sheet or an electrolytic tank [1]. The choice of an appropriate method will depend on the particular circumstances including the experience and knowledge of personnel available, complexity of the problem, the accessibility of hardware and software, and whether the solutions are for development, design or urgent investigative purposes. There is no doubt that in large organizations computer programs are applied extensively. Due to cost, however, the simpler techniques may still be preferred in some cases, especially for two-dimensional problems.

Programs for solving symmetrical 3-D problems by the Finite Element Method have been in use for many years. More recently the Boundary Element Method seems to have been developed for truly 3-D asymmetric conditions. In this technique a mesh is required only at the electrode boundaries and dielectric interfaces. This reduces the storage and memory needed and usually gives stress values directly as well as the equipotential distribution. The latter is often of value in the initial design of complex insulation systems. Large computers and costly programs involving many hours of running time are still necessary for the solution of practical problems such as the GIS components discussed, for example, in Reference 2.

Because a significant number of insulation failures are related to the electric stress distributions – magnitudes and spatial rates of change – consideration is now given to a number of common field conditions. Such analyses and associated PD/breakdown data can be helpful in interpreting the results of monitoring and testing of specific designs, including some complex configurations. Stress-enhancement factors – maximum stress ÷ average stress – have been determined for a number of simple electrode systems containing one dielectric, usually air or SF_6. A summary of a few of these factors is given in Reference 3.

3.2.1 Multiple dielectric systems

The simplest multiple dielectric configurations occur in uniform (parallel) and concentric fields.

3.2.1.1 Parallel electrodes

In Figure 3.1 is shown the condition where dielectrics are located between parallel electrodes. By neglecting the edge effects this arrangement can be used to determine the stress distribution in a parallel field in which several dielectrics are located in series, including liquid and gas layers as in laminated structures.

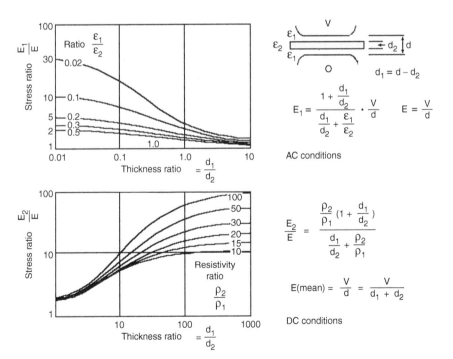

Figure 3.1   *Multiple dielectrics in a parallel field (variation of stress for two dielectrics (n = 2, d_1, d_2))*

If n dielectrics are represented then the stress in layer i is

$$E_i = \frac{V}{\varepsilon_i(\frac{d_1}{\varepsilon_1} + \frac{d_2}{\varepsilon_2} + \ldots + \frac{d_n}{\varepsilon_n})} \tag{3.1}$$

and the capacitance

$$C = 2\pi\varepsilon_0 \frac{1}{\frac{d_1}{\varepsilon_1} + \frac{d_2}{\varepsilon_2} + \ldots + \frac{d_n}{\varepsilon_n}} \text{ farads/metre}^2 \tag{3.2}$$

For the particular case of two dielectrics ($n = 2$) – for example gas (air)/porcelain, gas (air)/pressboard, gas (air)/resin, oil/pressboard (or paper), paper/resin – nondimensional curves giving the ratio of the stress (E_1) in dielectric n_1 and the average stress (E) may be calculated as in Figure 3.1. These cover the range of permittivity ratios of 0.02 to 0.5 and resistivity ratios of 100 to 10 for a range of dielectric thickness ratios of d_1/d_2. Applications of these simple but useful concepts may be demonstrated quantitatively as follows.

(a)   If the electrode spacing $d = d_1 + d_2 = 10$ mm, the air gap $d_1 = 3$ mm ($\varepsilon_1 = 1$), the barrier thickness $d_2 = 7$ mm ($\varepsilon_2 = 3.5$), the applied voltage V, then from (3.1) the air stress $E_1 = 0.2$ V/mm.

From Figure 2.2 the breakdown of a 3 mm-thick air gap in a uniform field is 2.59 kV/mm, thus the configuration would be expected to discharge when $E_1 = 2.59$ kV/mm or V $= 13$ kV(RMS). The calculation is an indication of how the introduction of a barrier of inappropriate thickness between HV and earth – or even between phases – may produce corona (partial discharges) in service for certain field conditions, thereby reducing the strength of the system. Failure by puncture may occur due to erosion of the surface by corona or, possibly, a flashover around the barrier even though the stress across the barrier is low – 0.74 kV/mm at corona inception in this case.

With a barrier of 4 mm thickness placed centrally between the electrodes the air space $d_1 = 6$ mm is divided into two 3 mm gaps and $d_2 = 4$ mm. The air stresses compute as 0.14 V/mm, giving a discharge inception voltage of 18.5 kV (RMS) assuming the stresses in the gaps are 2.59 kV/mm.

Although the inception voltage is increased by reducing the barrier thickness it is still below that of 21.9 kV for a gap of 10 mm without a barrier. The latter would be unstable with a near-uniform field containing particles, moisture and other pollutants as in a practical system – for example, bus-bar arrangements in metal-clad switchgear. Equipment test stresses would be chosen to be much lower than those deduced from the Paschen curves (see Figure 2.2), but the data are of value in choosing the optimum configuration and in analysing failures.

(b)   For the above condition, where the duct or gap is filled with oil ($\varepsilon_1 = 2.2$, 3 mm thick), the barrier is pressboard ($\varepsilon_2 = 4.4$, 7 mm thick), (3.1) gives the value of $E_1 = 0.153$ V/mm. From Figure 2.3 a 3 mm-thick oil gap would be expected to break down at approximately 14 kV/mm (RMS) corresponding to an applied voltage of V $= 91$ kV (RMS) with a stress of only 7 kV/mm in the solid and an average stress of 9.1 kV/mm (RMS). This average stress can be increased by reducing the barrier thickness and forming, for example, two 3 mm-thick ducts. By a procedure as in (a) the voltage for inception becomes 112 kV (RMS), although the gap still limits the overall strength. These stress values may not be achieved in practice in transformers where the allowable interwinding oil stress during short-term testing might be of the order of 10 kV/mm. This is well below the uniform field conditions represented in Figure 2.3. If the stress-v-volume graph in Figure 2.3 had been applied for a large volume, a lower voltage would be expected to produce partial discharges. Such considerations are taken into account when designing large insulation structures as in power transformers.

It will be noted that as $d_1 \rightarrow 0$ (or $d_1/d_2 \rightarrow 0$) the stress $E_1 \rightarrow E \cdot \varepsilon_2/\varepsilon_1$.

Similar calculations may be carried out for DC conditions in oil. In such cases the stress in the solid material will be higher than in the oil. For example, with a resistivity ratio of 100 (oil $- 10^{11}\Omega$m, impregnated paper/pressboard -10^{13} Ωm) and $d_1/d_2 = 1/9$ the stress in the oil is very small (see Figure 3.1). If the oil gap is relatively large the solid may be overstressed. During the application of direct voltage tests to AC equipment this condition must be recognized when choosing levels to be applied. With an air gap (high resistivity), partial discharges can easily be induced but are usually of a low repetition rate. The 'hissing' from high-voltage DC test sets and charging units is typical.

3.2.1.2   Concentric electrodes

Another simple configuration that can often be of assistance in interpreting test results is that of concentric cylinders, Figure 3.2 (a).

In this case the field distribution is logarithmic, the magnitudes depending on the radial location and the value of the permittivity (for AC conditions) or conductivity (for DC conditions). The stress at radius r_i is given by the following, where ε_x is the

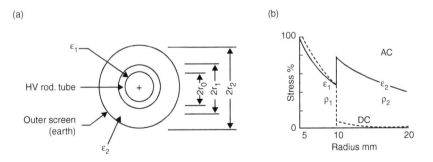

Figure 3.2   Multiple dielectrics in a concentric field. (a) Configuration for two dielectrics; (b) Variation of stress across two dielectrics

$$r_0 = 5\,mm, \quad r_1 = 10\,mm, \quad r_2 = 20\,mm. \quad \varepsilon_1 = 3.5, \quad \varepsilon_2 = 2.2$$
$$\rho_1 = 10^{13}\,\Omega m, \quad \rho_2 = 10^{11}\,\Omega m$$

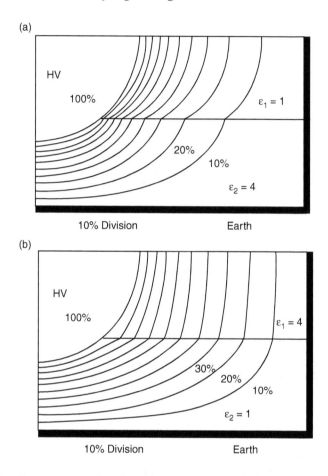

*Figure 3.4   Equipotential plots for partially embedded electrode. (a) $\varepsilon_1/\varepsilon_2 = 1/4$;
(b) $\varepsilon_1/\varepsilon_2 = 4/1$*

the ambient is very low. The concepts also apply if the ambient is a gas as in GIS, dry-type transformers, air-insulated switchgear and high-voltage machines.

The curves in Figure 3.5(b) for the main gap are based on permittivity, conductivity and spacing values for practical systems. The principles can be applied to other insulation arrangements incorporating different materials.

3.2.3   Multiple electrode configurations

During the investigation of certain types of fault it is often unclear as to why a particular tracking or flashover/puncture path was followed. Such a situation can arise where three or more electrodes are involved. Stressing in a well-insulated direction can produce PDs that may cause a failure along an unrelated path. This has been demonstrated experimentally at high voltage with the electrode system represented

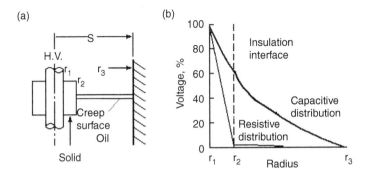

Figure 3.5   Insulation support for insulated high voltage lead. (a) Basic configuration; (b) Voltage distribution in main gap for AC and DC conditions (solid/oil system)

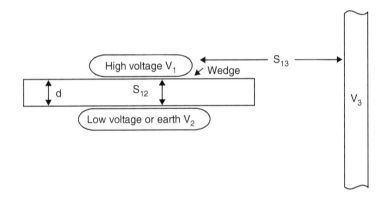

Figure 3.6   Effect of normal and tangential stressing in a multiple electrode system

in Figure 3.6. This is an example of a practical test in oil in which the flashover voltage strength along the insulation board was reduced by an increase in the stress value across the insulation thickness.

Although paths S_{12} and S_{13} may be considered independently, the initiation of PDs in the oil wedge due to the 'normal' stress between V_1 and V_2 can reduce the breakdown strength along the 'tangential' path S_{13} between the electrodes at voltages V_1 and V_3. Similar situations can occur in dry-type equipment. Such a scenario requires careful analysis of the electric field in relation to the original design and how this may have influenced the failure pattern. A more complex situation exists in transformer windings, where each turn may be treated as a separate electrode at a different potential from the others and from adjacent windings. In dry-type configurations, where discharges in the wedge may be induced more easily, such stress conditions can be very significant.

Simultaneous testing between phases and to ground in three-phase equipment may produce similar situations.

3.3 Electric stress control

In order to ensure the most economic use of space and materials, insulation designers have employed a range of techniques effectively to reduce the maximum to average stress ratio in the configurations utilized in high-voltage equipment. Under some conditions such methods may not be helpful. For example, during tests with insulated rods to a plane in oil it was found that the more uniform fields gave lower average stresses at breakdown. This was due to the 'volume' effect associated with the tendency for particles to be uniformly distributed throughout the oil space. In the case of the less uniform systems it can be postulated that the particles are attracted to the more highly stressed regions, thereby producing an overall oil gap of greater strength in series with the insulated high-voltage electrode.

The outstanding examples of stress control are in cable terminations, bushings, instrument transformers, insulator strings, stator bars, switchgear and power transformers. In many situations these insulation systems must be designed to withstand impulse voltages of twice or more the magnitude of the RMS value of the corresponding power frequency test voltage.

Many methods of stress control are practised as described in numerous papers and books. A number of areas of application of the techniques are summarized below and, where appropriate, reference made to diagrams in this book.

(a) Reduction of stress at high-voltage external terminations by
   • use of rings and similarly shaped electrodes in substations on support structures, surge arresters, circuit breakers, disconnect switches (isolators) and other devices (Figures 1.4–1.6);
   • application of continuous shields, or possibly 'cages', for bushing (Figure 5.11) and sealing end terminations.

(b) Reduction of stress at high-voltage conductors in air by
   • configuring two or four conductors in parallel as on transmission lines.

(c) Reduction of stress at internal support structures by
   • shaping of cast-resin spacers to improve the equipotential distribution at the conductors of SF_6 busbars;
   • dimensioning of insulation to reduce critical stress at HV lead supports (Figure 3.5).

(d) Reduction of stress at the edges or corners at earthed exits by
   • contouring earthed foils in bushings (contoured or part of capacitor foil system), and in sealing ends (Figures 4.9 and 4.12);
   • adding semiconducting tapes or similar compounds to plastic cables (Figure 4.12) and to stator bars (Figure 5.2).

(e) Reduction of stress external to oil-immersed transformer windings by
   • including metallized end rings in disc-type windings (Figure 5.10) and outer shields in layer-type windings (Figure 5.5(c)).

(f) Reduction and control of stress within the main structures of devices by
 • winding in foils within current transformers (Figure 4.4) and in capacitive type bushings (Figure 4.3);
 • adjusting the turns arrangement to give capacitive control of surge voltage stresses in transformer windings (Figure 5.4);
 • fitting of metallic shields around line-end tap changers;
 • inclusion of shields in through joints of high-voltage cables (Figures 4.10 and 4.13) and dry-type terminations.

Some of the above aspects will be highlighted in Chapters 4 and 5 during the description of possible faults in particular equipment and components.

3.4 Summary

The chapter reviews many of the factors involved in the design of electrical insulation as applied in power-system equipment – ranging from physical, mechanical and ageing aspects through to the determination of electric stress.

Stress analyses of simple insulation configurations are presented. Such concepts are helpful in the determination of the order of magnitude of electric stresses, perhaps after a failure on test or in service, and in choosing the most appropriate computer-based program to investigate details. These basic analyses also assist in understanding the adverse conditions that can be present or arise. A number of electric stress-control methods as applied in practice are listed at the end of the chapter.

An appreciation of the problems associated with the insulation design of particular power equipment is desirable when choosing or developing a condition monitoring system.

3.5 References

1. Vitkovich, D., *Field analysis: Experimental and Computational Methods* (D. Van Nostrand, New York, 1966)
2. deKock, N., Mendik, M., Andjelic, Z., and Blaczczyk, A., 'Application of the 3D boundary element method in the design of EHV GIS components', *IEEE DEIS Electrical Insulation Magazine*, May/June 1998;**14**(3):17–22
3. Ryan, H.M. (ed.), *High Voltage Engineering and Testing*, 2nd edn (IET, London, 2001)
4. Mason, J.H., 'Discharges', *IEEE Transactions on Electrical Insulation*, August 1978; **EI13**:211–38
5. Kreuger F. H., *Partial Discharge Detection in High-Voltage Equipment* (Butterworth & Co., London, 1989)
6. Binns, D.F., and Randall, T.J., 'Calculation of potential gradients for a dielectric slab placed between a sphere and plane', *Proc. Inst. Electr. Eng.*, 1967;**114**:1521–8

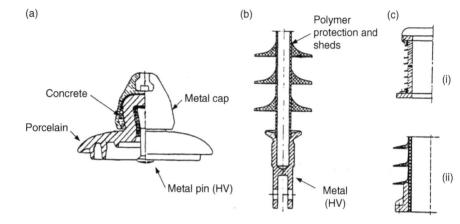

Figure 4.1   Examples of suspension and post insulators. (a) Porcelain suspension insulator; (b) Composite polymer suspension insulator; (c) Insulator shells (i) porcelain (ii) composite

of 'dry bands' on the porcelain surface. The bands are created by the electric stress in localized regions on the surface being of sufficient magnitude to produce losses in the pollutant, which may dry out because the ambient condition (e.g. the moisture evaporation rate) does not allow dissipation of the heat. This can result in a build-up of voltage across the band with subsequent discharges or arcing [2].

Another fault that may or may not be troublesome with ceramic insulators is the presence of partial discharging (corona). Although the PDs, usually associated with the metal work, rarely cause damage to the materials, they can result in unacceptable radio and TV interference.

Failures of insulators were caused by excessive deposition of ice and snow during storms in the USA and Canada. These types of failure are surveyed in a 2005 CIGRE report [3], which contains a number of references related to the condition. EPRI investigations have indicated that a number of cap-and-pin insulators would be expected to lose much of their insulating properties after 60 years in normal service [4]. Despite the excellent performance of the older units, cases have been reported of catastrophic mechanical failures of recently manufactured ceramic insulators.

Long-rod composite polymeric insulators are increasingly being applied despite the naturally cautious approach of utilities. The advantages include reduction in weight – giving easier handling – and somewhat improved performance under heavy pollution conditions. However there is still some concern regarding chemical changes due to dry-band-arcing, corona and weathering. Laboratory and field tests have been carried out for many years but there appears to be a wide range of opinion and experience as to the long-term effects. In the United States a high percentage of all new insulators are composites, while in other countries the introduction of polymer units is being carried out with caution or possibly on a trial basis.

The principle of the construction of long-rod composite insulators is shown in Figure 4.1(b) – a stress ring at the lower end is not included. The load-bearing inner rod consists of fibreglass-reinforced polymer (FRP) – epoxy resin probably being the most used polymer – terminated with metal ends. These are shaped to minimize corona, whose presence can contribute to brittleness and cracking of the glass–resin composite. The polymer seal must be maintained in order to prevent access of moisture. The complex processes associated with 'brittle' failures have been studied for many years. Part of the research in the USA supported by the Electric Power Research Institute (EPRI) is reported in some detail in Reference 5. The catastrophic failures develop radially across the glass fibres, probably in the high-voltage region, and appear to be due to the combined influence of the working mechanical stress, acidic attack due to the interaction of the materials (resin and glass), the electric stress, corona effects (ozone) and moisture ingress/diffusion. The use of polyester resin is considered to be unsuitable by the EPRI investigators. Although the number of failures is low (200) compared with those now in service ($> 10^6$) [5] the long-term effects are not known. However, various design and manufacturing changes have been made based on the studies so far, including recommendations regarding correct handling procedures.

In composite insulators, the weather sheds are formed over the rod and usually consist of silicone rubber (SIR). Chemical bonding is effected between the sheds and the rod and between the sheds themselves. As the electric stress acts along the rod fibres and the longitudinal interface, it might be expected that any manufacturing or service problems in these regions could result in electrical as well as mechanical weaknesses [4].

SIR is probably the preferred material at the present time, because its hydrophobic properties (minimum affinity for water) are good and it has a higher resistance to UV radiation than some other materials and also to dry-band arcing. After exposure to intense discharges, SIR forms silica, which is non-conducting. Other materials that have been used and are in service include ethylene propylene rubber (EPR) and cycloaliphatic epoxy.

For many years room-temperature-vulcanized (RTV) SIR high-voltage coatings have been used by some utilities for improving the pollution performance of porcelain insulators. The behaviour of such systems is discussed by Cherney *et al.* [6], where it is emphasized that, unlike the case with porcelain, low-energy corona from end fittings can degrade the RTV–SIR coating. This can be a significant problem if the original source of corona is not removed by modification of rods, end rings and similar.

The monitoring of potential faults associated with insulators may be detected by thermal (or acoustic) scanning, as discussed in Chapter 9.

4.1.2   Post insulators

Post insulators may consist of a solid core, a shell of porcelain or, in the case of composites, an inner, fibre-wound epoxy cylinder with metal terminations and a protective polymer covering. For the last of these, sheds are provided of a compatible material – e.g. silicone rubber. Identical faults to those for the suspension units have not been reported [4], although it is possible that high crimping pressures of the terminations may have caused apparently similar mechanical failures at the earthed end, as well as at the high-voltage connection. Larger-size shells are also used in bushings, instrument transformers and surge arresters.

4.2   High-voltage bushings

Bushings in their various forms are used wherever it is necessary to pass a live conductor through a metal plate or tank at a different potential – usually at earth potential. The basic requirements are indicated in Figure 4.2(a). A bushing consists of a metal rod (or tube) supported by a simple insulation structure at the lower voltages through to higher-voltage units containing many stress-grading foils within the insulation. Commercial bushings are classified in Standards such as IEC 60 137 [S4/1].

The method of improving stress control by means of isolated conducting foils is shown in Figure 4.2 (b), where a comparison is made between an ungraded and graded configuration of a hypothetical bushing. It is apparent that the voltage distributions axially along the insulation surfaces and radially through the structure are much more uniform in condition (ii) than in the uncontrolled case (i).

In practice it is not possible to achieve the ideal of uniform distribution in both axes by adjusting the relative capacitances [7]. Computer programs are utilized for solving the complex calculations required to determine the number and dimensions of foils necessary to give acceptable stresses within the structure. The values are related to the puncture and creep strengths of the various solid insulations and the associated

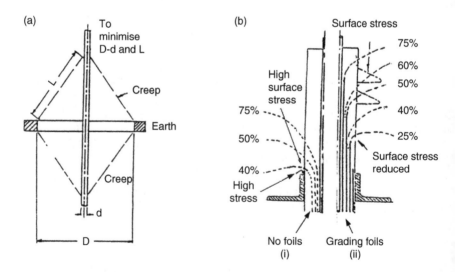

Figure 4.2   Basis of high voltage bushings. (a) Design requirements; (b) Equipotential distribution (i) No foils (ii) Grading foils

ambient medium of air, SF_6, or oil. According to the voltage class the number of foils varies from zero at 11 kV to two or three at 33/66 kV through to perhaps one hundred or more at the highest levels. Care must be taken that overstressing does not occur at the foil terminations. The outer foil of the majority of high-voltage bushings of 72 kV class and above may be 'floated' in order to provide a means of monitoring the insulation condition in the factory and in service. It is normally earthed. If not earthed, or connected through an appropriate impedance, the foil will rise to a high voltage and probably cause an insulation failure.

The types and forms of the insulating materials in bushings depend on the manufacturer's practice, the voltage class and, to some extent, the age of the product. Older units were insulated with SRBP and the solid structure immersed in oil within a porcelain shell. Such designs can operate successfully up to 300 kV but there have been problems regarding partial discharging within the resin paper wraps. Both loss angle and PD measurements are applied to assess the state of, for example, SRBP 66 kV and 132 kV bushings, many of which of the older units have internally eroded insulation, often produced over long periods of time. In some cases replacement programmes of up to 30-year-old units have been instituted based on such measurements [8]. Some of these bushings have been operating with PD levels of several hundred pC. High-voltage resin-vacuum-impregnated paper bushings have been produced for a number of years including units up to 500 kV. However, it appears that the oil-impregnated paper type is still preferred for many applications at the higher voltages.

Oil-impregnated paper bushings are built into, for example, power transformers, shunt and series reactors, switchgear and enclosures at substations up to the highest voltages. In these designs it is essential that the multi-foil paper structure be well dried and efficiently impregnated with dry oil under low vacuum when mounted in the porcelain housing (shell). More recently, composite shells – possibly with silicone sheds and protection – have been manufactured. Low partial-discharge levels must be achieved in all high-voltage bushings in order to prevent problems in service. Also low values are necessary in any bushing type installed in equipment in which PD levels of tens of pCs or less are required to be measured. A state of dryness must be maintained during operation, otherwise the dielectric losses may rise significantly, resulting in thermal deterioration of the insulation and possibly the initiation of high-level PDs (hundreds/thousands of pCs). Both changes can result in failure and internal explosion. Such events have resulted in intense fires and widespread damage within large power transformers. The application of composite shells reduces the damage and danger to personnel should an explosion occur.

An example of external flashovers of wall bushings due to salt pollution is described in Reference 9. The failure resulted in the outage of a complete power station. The bushings were replaced with another design and a leakage-current-monitoring system installed having an initial sensitivity level of 13 mA. Automatic washing equipment was also installed as a trial.

Typical equipotential distributions around the oil end of a conventional and a re-entrant high-voltage transformer bushing mounted in a turret are depicted in Figure 4.3 [10]. The manner in which the field is controlled by the capacitive grading (foils) and the stress distributor at the point of entry of the high-voltage lead is apparent for

However, many thousands of existing oil–paper-insulated instrument transformers will continue to operate for many years to come [8]. It is important that users be aware of possible failure scenarios, especially for the oil-impregnated units, and of the site-monitoring methods now available or under development (see Chapter 9).

4.3.1   Oil-impregnated current transformers

The constructions of HVCTs vary considerably, as indicated in Figure 4.4. These may be designated (a) hairpin, (b) eye bolt and (c) bar primary. Earlier designs were reviewed in a report prepared for Study Committee SC 23 of CIGRE by WG 23.08.

In cases (a) and (b) the oil–paper insulation structures are equivalent to electro-statically graded bushings with the inner electrode at high voltage and the outer foil at 'earth', adjacent to the earthed core around which are located the secondary low-voltage windings. A 330 kV CT might incorporate 30 main foils and a 66 kV unit, only two or three, each with end rings. In some very high-voltage designs short intermediate foils may be used between the ends of the main foils in order to improve the end stresses without incorporating local stress rings. The structures are enclosed in a housing mounted on a metal tank, the whole being dried at temperatures of up to 120°C and vacuums of less than 0.02 torr and then filled with processed oil under vacuum. Because of the large diameters required, each leg of some hairpin designs is accommodated in its own housing. It is noted that in Japan a porcelain bushing housing (shell) of 1.5 metre diameter and 11.5 metres length has been developed for 1 000 kV AC operation but such sizes are very expensive.

The bar primary arrangement (c) is, in effect, an inverted bushing with the inner electrode tube containing the connections from the secondary windings, which are mounted on the earthed core, also insulated from the high-voltage primary bar at the line end. The major internal insulation from the high-voltage outer to the inner

(a)   (b)   (c)

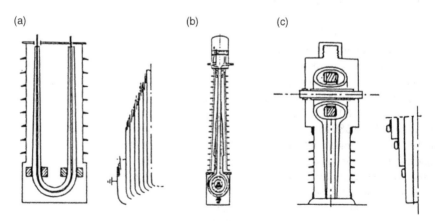

Figure 4.4   Examples of current transformer (HVCT) configurations. (a) Hairpin type; (b) Eye bolt type; (c) Bar primary type

grounded tube is provided by an axially and radially voltage-graded insulation system. A housing is provided for containment of the oil and for supporting the overall structure. The head of the type in (c) may be metal or of resin. Such units are available for operation in 765 kV systems. The capacitances of the various HVCTs are in the range of a few hundred to possibly a thousand picofarads.

Most of the major HVCT failures have been associated with paper/oil hairpin units. The breakdown modes are difficult to identify but probably involve one or more of the following:

(a) Degradation of the paper system due to ingress of moisture through leaking seals. A fault could develop as the dielectric losses increase, resulting in thermal instability at the operating stresses of 3–4 kV/mm. This, together with surge effects, may explain failures after relatively short periods of 10 to 15 years.

(b) Development of trapped gas in the body of the insulation during the process in (a) resulting in the inception of partial discharges, initially of low levels and increasing to values of thousands of pCs before failure by puncture. 'X-wax' and small black spots in non-failed regions are indicative of long-term thermal instability followed by partial discharging. Such effects may arise due to poor dry out and impregnation during manufacture.

(c) Overstressing at a foil termination produced by deterioration or physical damage could result in PDs leading to puncture.

(d) Inconsistencies in the taped paper structure allowing partial discharges to develop due to reduction in thickness and, also, the creation of oil gaps (see Figure 4.5).

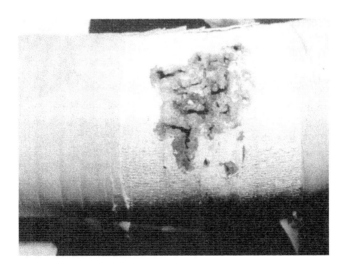

Figure 4.5   Failure of HV current transformer oil–paper taping due to dielectric overstressing and partial discharges [26] [reproduced by permission of CIGRE]

(e)   Long-term disintegration of the connection between the internal DDF terminal (normally earthed) and the first foil used for the measurement. This may produce gas, which results in axial flashover through the oil. The fault could be due to a cumulative effect produced by the surge currents and associated arcing created during switching of local air isolators. An optional special test is included in the revision of the instrument transformer standard [S4/2, Annex B] requiring the application of a minimum of 100 chopped waves.
(f)   Bad connections or floating metal at the DDF tap, which may result in a condition as in (a) (Figure 2.8(vii)).

Condition (d) may explain some failures after less than a year's service. The failure mode probably involves paper burning between several foils and the production of PDs – possibly increasing to ten nCs or more near failure. A common feature seems to be that the final scenario develops quickly – perhaps in days or even hours – making prediction of breakdown by means of the commonest monitoring technique of periodic oil sampling of DGA ineffective. Also, it has been found in some cases that a significant change in the DDF value would not necessarily arise if, for example, considerable burning between two foils was present [13]. In other instances low partial-discharge values have been recorded, even though tracking was established. Nevertheless, continuous monitoring of DDF and, where practical, PDs is highly desirable for critical installations. It is noted that in these types of failure a high internal pressure is built up due to arc gases being produced within the small volume of the CT housing. This can, and does, result in explosions, causing the shattering of the porcelain with dangerous consequences. Other anomalies are described in Reference 13.

Partial-discharge levels may range from hundreds to many thousands of pCs in a 'dielectric stress' type failure. DDF values possibly increase to hundreds of mRs in a 'thermal runaway' breakdown mode.

4.3.2   Dry-type current transformers

In the surveys quoted above it was reported that major failures of cast-resin CTs in the voltage range $>60\,kV$ to $<200\,kV$ were 0.175 per cent of their population. It is not known whether there is a dominant cause for these failures. However, with such components good quality control is essential in order to ensure that no voids are produced during processing. An example of such a problem was a $33\,kV/\sqrt{3}$ bar primary unit that, on receipt from a reputable manufacturer, was found to have a PD level of 200 pC at operating voltage. Due to operational pressures the CT was installed and failed after one year's service. Pre-service and, perhaps, periodic PD measurements may be justified for this type of construction. Another example of a poor-quality product was an 11 kV CT designed for mounting on a bushing. In this case a low-quality plastic insulating tube and associated compounds were used, resulting in an unacceptably high DDF value. The excessive dielectric losses could have produced a service failure. Again, the unit was from an established manufacturer.

At the higher voltages, live tank dry-type units are now common, utilising SF$_6$ gas as the insulant between the primary bar and the earthed components located inside the head. The support structure may be of porcelain or polymeric material. An appropriate gas pressure must be maintained during the life of the CT and its level monitored in service.

4.3.3   Capacitor-type voltage transformers – CVT

There are two basic forms of voltage transformer for use at high voltage: the capacitor divider type (see Figure 4.6) and the magnetic type – MVT (see Chapter 5).

The cheaper, but possibly less accurate, capacitor-type (CVT) design is effectively a voltage divider with a tuned low-voltage arm. The voltage across this arm would be approximately 19 kV with a capacitance ratio of, for example, 3 800/38 000 pF for a 330 kV unit and, possibly, 12 kV (5 600/30 000 pF) for a 132 kV CT. The capacitors consist of many series elements subjected to a slight positive pressure with a steady-state voltage of, typically, 2 kV across each. The design of the elements and the operating stresses (20 kV/mm) are similar to those used in paper/oil power capacitors. Where paper is included, the insulation system is dried and vacuum impregnated with processed oil.

In the case of the capacitor units (Figure 4.6) major failures have been related to discharging at the end of foils of individual elements (Figure 3.3), thermal runaway and oil leakage exposing the upper elements. It is possible for two or three elements to be shorted and a major failure not to occur immediately. Some operators consider that physical inspection, e.g. for oil leakage, can be a very useful indicator of the internal state of an instrument transformer. Such a procedure is labour-intensive and may be hazardous in the case of suspect units. Because the loss of elements will vary the output voltage, the inspection of CVTs can be minimized by continuously monitoring any small changes in the output voltage.

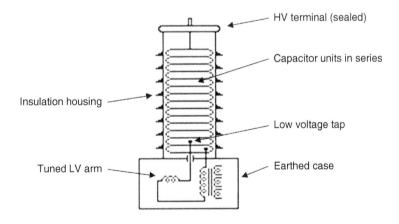

Figure 4.6   Capacitor voltage transformer (CVT) configuration

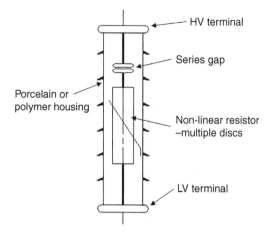

Figure 4.8   Layout of a single-stage gap-type surge arrester

(e) mechanical fractures in the metal-oxide material (gapless type) due to thermal runaway after a high current surge; also damage due to surge current concentration at the edge of the electrode can result in failure [15];

(f) resultant damage to the discs created by previous multiple-stroke lightning surges; this is a condition not covered by routine or type tests and may be very significant [16].

Metal-oxide arresters are inherently faster-acting than the gapped type, since there is no time delay due to series air gaps extinguishing the current. It seems that the metal-oxide units are proving more reliable than the previous designs.

In critical locations, e.g. adjacent to a major transmission or generator transformer, it may be justifiable to monitor arrester conditions continuously – see Chapter 9.

4.6   High-voltage circuit breakers

There are many types of circuit breaker installed in power supply systems. These range from 11 kV air break metal-clad or resin-insulated systems, bulk oil units from 11 kV up to 275 kV, low-oil-content live tank designs, air blast of dead and live tank configurations up to 400 kV and, in recent years, the widespread application of SF_6 units. Different designs of the last of these are applied over the whole voltage range. The majority of new breakers are of the SF_6 type and include live and dead tank types. Examples of SF_6 units produced by six manufacturers are presented in Chapter 8 of Reference 18. However, many of the existing units at the lower voltages are oil-filled and a number of air-blast types remain in service at 132 kV and possibly 275 kV, all of which will need regular periodic inspections until replaced.

For some applications 11 kV and 33 kV vacuum interrupters are installed. The switches are vacuum-sealed and should not require maintenance throughout the specified life. These breakers can produce steep transients, as can SF_6 circuit breakers at the higher voltages. Protection of the circuits and connected plant can often be effected by correct use of metal-oxide arresters (MOA).

Insulation failures in circuit breakers would be expected to be associated with deterioration of the arc-suppressing medium or the insulation of components between high voltage and ground. According to the age and type of breaker, the components might include bushings, laminated wooden operating arms in older designs, SRBP and resin-impregnated glass-fibre operating rods, tubes of resin-impregnated glass fibre for compressed-air or SF_6 gas feeds in live tank configurations, outer shells and support insulators of porcelain or polymers.

Insulation breakdowns within SF_6 circuit breakers can be due to chemical by-products producing conductive areas on insulator surfaces, protrusions on conductors, conductive particles, faults in the solid dielectrics including voids, trapped moisture and poor adhesion of parts. Dielectric overheating produced by bad contact in the current circuit can also lead to breakdown. In practice, failure rates due to insulation deterioration are low compared with other factors associated with breaker operations.

Internal flashover due to tracking along laminated components in older bulk oil breakers can result in the production of gases and subsequent explosion and fire [17].

An interesting example of a failure to ground was due to a wooden operating rod within an oil-filled porcelain shell supporting the live-end tank of a 330 kV breaker. Due to moisture ingress and the incorrect location of the rod with respect to the shell, it was concluded that flashover had probably occurred from the high-voltage termination over the outer porcelain surface to an intermediate metal flange, across the reduced oil gap to the rod and then along the rod–oil interface to ground. In addition to visual examination the probable sequence of events was deduced from measurements of the insulation resistance (IR) values of the various paths involved.

Another case involving moisture ingress measurements on laminated components from oil-filled switchgear enabled a criterion based on IR values to be determined for the continued operation of similar equipment – thus acting as a simple condition assessment for the particular insulation system [18].

The application of unsatisfactory insulation as spacers between busbars in metal clad 11 kV switchgear created major problems in a number of substations. Following an inter-phase failure at one site, diagnostic techniques, including DDF and PD measurements with ultrasonic detectors, were applied. These procedures allowed surveillance of operating substations while others were being modified. Several years were required to update the many substations involved. The success of the project showed the value of having available a range of monitoring systems together with the necessary specialized engineers and test personnel.

The application of partial discharge testing in an open, conventional, high-voltage switchyard is often restricted by the presence of air corona from the bus-bar system. Obvious sources are rod gaps and faulty insulators. Site experience has shown that many designs of air isolators used for protection during maintenance procedures have sharp-edged guides for mechanically locating the horizontal arm as it closes onto the line. Corona from these guides can be easily identified using a handheld ultrasonic detector and may be of several thousand pCs magnitude. In many cases it would appear possible to reduce the disturbances by modification of the configurations to allow measurements on adjacent equipment. Other corona sources noted included the corners of lugs provided for temporary earthing purposes on 132 kV busbars and, in one case, the lack of bonding between a 66 kV bar and the metal cap of the supporting insulator. This fault produced a disturbance of greater than 10,000 pC, preventing the completion of any useful PD measurements in the substation. Experience has shown that a 6 mm-diameter rod mounted on a 132 kV bushing (76 kV to ground) will produce a PD of 1 000–2 000 pC, depending on the weather conditions.

Routine periodic monitoring (DDF and IR) of circuit-breaker insulation would probably include the bushings and any associated busbars as appropriate. Under some conditions it may be possible to carry out online PD measurements utilising a high-voltage reference capacitor (see Chapter 9) or high-frequency probes (capacitive or current transformer – Chapter 8). By application of a separate source supply and a differential circuit measurement, sensitivities of 10 pC or less have been reported during site tests in high-voltage switchyards.

4.7   Gas-insulated systems (GIS)

In order to overcome the space limitations, and some environmental problems created by conventional open-type substations, gas-insulated systems have been developed in which the circuit breakers, other switchgear, disconnectors (isolators), instrument transformers, surge arresters and busbars are all enclosed in metal earthed chambers. Epoxy insulators are used for supporting the structure in the compressed SF_6 gas. The systems are designed to keep the gas dry and oxygen-free, thus minimising long-term insulation deterioration. Since no maintenance is required except for inspections after a specified number of switching operations, GIS are normally considered to be maintenance-free. Complete systems are available for voltages up to at least 765 kV. An earlier 132 kV design was installed in Sydney, Australia, in the 1970s for connecting the outputs of 330/132 kV transformers to the local distribution system. The conventional transformers were supplied from outside the city through 330 kV single-phase cables. Much information is available regarding the design and performance of GIS substations [18].

Special diagnostic measurement techniques for GIS have been developed over a long period of time but are not yet standardized (see Chapters 6 and 9). The majority of the methods are applied online and include the electrical detection of low-level PDs of less than 10 pC at very high frequencies and by ultrasonic systems. Low levels are also required at overvoltages during commissioning following assembly of the complex structures. Because of the relatively low capacitances, separate-source PD tests are practical. By-products detected during analysis of SF_6 gas samples may represent a fault as well as indicating the possibility of damage to the insulation and other materials in the chambers. The major causes of deterioration, or flashover, are associated with the presence of small metal particles, loose bolts, bad contacts, faults developing in the epoxy spacers due to decomposition of the gas, and perhaps voids that were undetectable during commissioning. Inappropriate electric field distributions along spacers may occasionally produce surface-creep problems. Detailed computer programs have been developed by manufacturers to achieve optimum electric field conditions around the electrodes and along insulation surfaces.

4.8 High-voltage cables

Much information is available describing the many designs and insulation systems of high-voltage cables and accessories. Electrically, mechanically and thermally, the configurations are simple concentric cylinders, except for the older, lower-voltage, three-phase belted cables, where no outer screen was provided for each core. When practical, the three screened cores of a three-phase system are contained in a common cable but, at the higher ratings and voltages, it is necessary to manufacture and lay them separately.

The electric field in the body of the cable follows that given by Equation 3.3 for one dielectric and that at the end by the type of distribution in Figure 3.3(a), where very high stressing occurs at the edge of the outer sheath. In this region stress control is effected by use of one of the many terminating methods available, the components of which are critical in the design and assembly of a cable system.

The many types of high-voltage cable require highly specialized production and installation techniques. At the present time, in respect of insulation, there are three major classes of cable in service.

4.8.1 Oil–paper cables

Oil–paper cables may be classified according to the following:

a) Older oil–paper 'solid' systems filled with oil/rosin – operating up to 33 kV and possibly 66 kV;
b) Oil–paper systems impregnated with low-viscosity oil at high vacuums and sealed in service under positive pressure – installed up to the highest voltages and ratings (hollow conductors are utilized for cooling in single-phase cables and channels or a pipe-type enclosure in three-phase designs);
c) Oil–paper system but filled with dry nitrogen under pressure;
d) Oil–polypropylene–paper (PPLP) systems for very high-voltage applications – this is a relatively new technology [19].

4.8.1.1 Faults in the bodies of oil–paper cables

(a) Sheath damage – accidental or due to chemical erosion – allowing the ingress of moisture, and in pressure cables the loss of oil, with subsequent degradation of the dielectric. Erosion problems would appear likely in distribution cables some of which are more than 40 years old.
(b) Thermal runaway caused by overloading beyond the specified design limits or the lack of adequate heat dissipation around the cable. The latter is a major factor in the installation of underground cables. The failures in Auckland in

Where a solid insulating material of different conductivity to the paper is used in such joints – and in other high-voltage components – the design must allow for the changed field distribution if direct-voltage tests are to be applied in the factory and during commissioning of the AC cable. Previous damage was suspected due to such a test following the failure of a 132 kV cable joint in service some years ago.

4.8.2   Extruded cables

During the past fifty years or more, extensive research into the development of plastic materials has resulted in their successful application as insulation for cables – initially for low-voltage wiring (e.g. PVC coverings) through to modern systems of XLPE, some of which are operating at 500 kV (AC). Test requirements for the various voltage levels are given in the Standards [S4/7–S4/9]. The widest application of extruded cables has been for medium to high voltages in the distribution field: ranging from 11 kV through to 33 kV/66 kV and a few systems at 132 kV and 500 kV. The production techniques have varied but the commonest, based on much experience, is the triple-extrusion method. The construction of a typical single core 22 kV cable is depicted in Figure 4.11.

During the process an inner semiconductive layer, the major insulation and an outer semiconductive layer are extruded simultaneously, thus producing, ideally, a good concentric field. In earlier methods the outer layer was added separately, which sometimes created voids at the interface and subsequent failure due to partial discharges. XLPE in particular is very susceptible to low-level PDs. The usual insulation is XLPE, although EPR is favoured by some manufacturers for specific applications.

Apart from avoiding the formation of voids during the manufacturing process, a major problem with XLPE has been the tendency for water trees to develop in the material. The effect has been minimized by replacing the earlier pressurized steam cross-linking techniques with, for example, the siloxane process or the continuous-gas-vulcanization dry-curing method in pressurized nitrogen and, when considered necessary in service, the use of a water barrier on the outside of the cable. It is noted that EPR and tree-retardant cross-linked polyethylene (TRXLPE) appear to be the preferred materials for medium-voltage (15–35 kV) cables in the USA [23]. The application does not require the use of water barriers. However, for 69 kV and above, barriers are often specified.

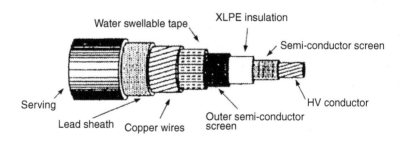

Figure 4.11   Construction of a typical XLPE 22 kV (RMS) single core cable

XLPE and EPR cables are now very competitive with oil–paper types at the medium-voltage range for a number of reasons including the possibility of operating at somewhat higher temperatures. At the higher voltages the application of plastic cable designs is increasing.

Although the manufacturing and acceptance tests for plastic cables are very stringent, a number of potential faults are possible in service. These include the following in the bulk of the cable.

(a) The development of carbonized trees in the main insulation due to voids or particles introduced by incorrect manufacturing. The initial faults may not have been detectable during the factory PD measurements.

(b) Localized separation between the main insulation and semiconductor outer layers resulting in PDs, possibly formed after manufacture due to lack of bonding and accentuated by bending of the cable.

(c) The creation of water trees in XLPE formed during manufacture – but now less likely – and in service following ingress of moisture. Failures in the older cables were relatively frequent due to this cause [23]. The 'trees' cannot be detected easily in the final product and become apparent only when almost bridging the dielectric. Advanced detection techniques are being developed including dielectric response measurements (see Chapter 6).

4.8.2.1 Terminations and joints for extruded cables

Many forms of terminations and joints have been developed for extruded polymeric cables. Some of these are described in principle in the report by WG 21.06 of CIGRE [24]. This includes outlines of techniques for transition joints for connecting oil-filled and polymeric cables.

One of the special problems associated with polymeric terminations and joints is the possible creation of faults at the interface between the cable insulation and the electrical stress-relieving material or semiconductor cone connected to the earth screen. The objective is to improve the longitudinal (axial) stress along the interface and the cable insulation. The examples in Figure 4.12 indicate how this may be effected by application of (a) a heat-shrink stress control sleeve, or (b) a semiconducting stress cone, or (c) a semiconducting cone with a conical thermoset casting and compression springs. At the higher voltages the longitudinal stress can be improved by winding layers of conducting film interleaved with polymeric film around the cable insulation to form a capacitor-type bushing construction [24].

The necessity for very good physical contact between interface surfaces is analysed in CIGRE Brochure 210 [25]. Among the findings, it is concluded that (i) the surfaces must be sufficiently smooth to reduce void sizes to a micrometre or so and (ii) sufficient pressure be applied by – for example – heat shrinking (Figure 4.12a), the elasticity of the rubber material or spring pressure forcing a rubber insert against an epoxy conical casting (Figure 4.12c). The mechanical and thermal characteristics must be such that no significant gaps or voids are formed when installing on site or due to movement in service. Although the maximum electrical stress will be radial, it is the longitudinal component that may cause failure. This is especially the case if

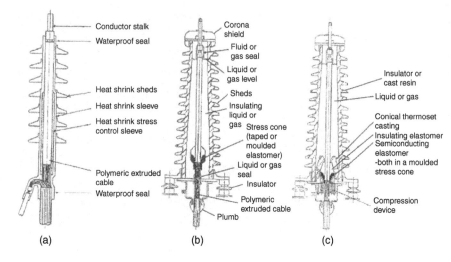

(a)   (b)   (c)

Figure 4.12   Possible outdoor terminations for polymeric extruded cables [24] [reproduced by permission of CIGRE]

voids have been formed and the radial stress value is sufficiently high to induce partial discharges. If the magnitude of the latter is of the order of only a few picocoulombs this might lead to failure within a few hours in some joint structures [25]. Outdoor terminations may have lower stresses and similar (or higher) PD values but failure would be expected to occur in the longer term. In order to minimize possible problems, emphasis is given to careful preparation of the surfaces, the precise following of the accessory manufacturer's instructions and the employment of skilled jointers.

In older designs a failure due to arcing can occur if the semiconductor elastomer of a cone is incorrectly assembled over the earthed screen and is not making contact. Such a condition has resulted in the semiconductor rising in potential above earth with subsequent discharge to the screen and failure.

The principles discussed above for terminations are applicable to the numerous forms of joints [24, 25]. In many respects the installation and the long-term condition of the joints are more critical because of their complexity and difficult access. A particular design of joint is shown in Figure 4.13 [24]. In Reference 26 the separation or relaxation of the stress-control materials in service was found to be a significant cause of failure.

Much attention is being given to the development of techniques for the condition monitoring of extruded cables and, in particular, their accessories (Chapter 9).

4.8.2.2   Gas-insulated transmission lines (GIL)

Gas-insulated transmission lines have been developed over many years for installation in locations where access, for example, to a city centre by a high-voltage supply is required. The lines comprise a rigid metal outer tube with an inner conductor mounted on specially designed epoxy resin insulators. The lines can be buried, above ground

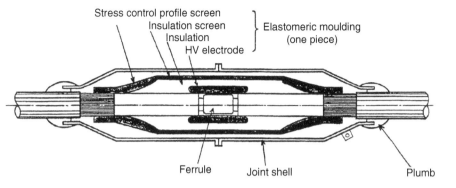

Stress control profile screen
Insulation screen
Insulation
HV electrode
Elastomeric moulding (one piece)
Ferrule   Joint shell   Plumb

Figure 4.13   Example of dry-type pre-moulded joint [24] [reproduced by permission of CIGRE]

or mounted in existing/new tunnels. An overview is presented in *Electra* [27], in which systems of up to 550 kV and 3 km in length are described. The gas mixture used is quoted as 80 per cent N_2, SF_6 20 per cent at a pressure of 7 bar, thus reducing the quantity of SF_6 required.

In an earlier 275 kV system of 3.3 km length the diameter of the outer aluminium tube is 420 mm and that of the high-voltage conductor tube 170 mm with an SF_6 gas pressure of 0.44 MPa. These systems would be expected to require minimal maintenance. The commissioning checks include partial-discharge measurements, probably utilizing UHF technology [27] (Chapter 9).

4.9   Summary

The chapter reviews how the various insulating materials are applied in a range of power-system equipment excluding rotating machines and transformers. Although the number of examples of possible failures is limited, those chosen indicate where problems might arise. Knowledge and understanding of the behaviour of new and aged insulation structures is essential when considering the need or otherwise for a particular condition-monitoring system. The values of the partial discharges expected for the different operating conditions give an indication of the sensitivity required for the condition monitoring systems.

4.10   References

1. Insulation News and Market Report, 1998;**6**(1):54
2. Williams, D.L., Haddad, A., Rowlands, A.R., Young, H.M., and Waters R.T., 'Formation of dry bands in clean fog on polluted insulators', *IEEE Trans. on Dielectrics and Electrical Insulation*, October 1999;**6**(5):724–51

S4/8 IEC 60840 (2004): Power cables with extruded insulation and their accessories for rated voltages above 30 kV (Um = 36 kV) up to 150 kV (Um = 170 kV) – Test methods and requirements (AS1429.2:1998)
S4/9 IEC 62067 (2001): Power cables with extruded insulation and their accessories for rated voltages above 150 kV (Um = 170 kV) up to 500 kV (Um = 550 kV) – Test methods and requirements

4.12   Problems

1. Capacitive grading is to be included in an oil impregnated bushing ($\epsilon_r = 3.5$). The details are as follows:

    Length of foil adjacent to HV conductor = 400 mm (l_1).
    The radial spacings between conductor/foils/earthed flange are 6 mm.
    The conductor radius (r_0) = 20 mm. The earthed flange inner radius (r_4) = 44 mm.
    The maximum radial stresses in the OIP of each layer are to be equal.
    Determine

    (i)   the foil lengths ($l_1 \ldots \ldots l_4$)
    (ii)   the capacitance values between foils ($C_1 \ldots \ldots C_4$) and that between HV and earth (C)
    (iii)   the maximum radial stresses
    (iv)   the maximum radial stress without foils

    The test voltage is 60 kV (RMS)
    Assume $Q = CV = C_1(V_0 - V_1) = C_2(V_1 - V_2) = \ldots \ldots = C_4(V_3 - V_4)$.

2. What are the limitations when measuring the partial-discharge characteristics of an AC extruded power cable on site (a) using a separate-source high-voltage supply and (b) under operating conditions? Assume a termination and a joint are in the section under test. Refer to Chapters 6, 7 and 9.

3. An extruded cable is terminated with a semiconducting stress cone. By assuming a radial operating stress at the cone and cable insulation of 2 kV/mm (RMS) calculate the thickness of a void/air gap within the interface that might discharge under this condition. Also, determine the maximum gap thickness allowable to avoid partial discharges during a test voltage application of $1.5U_0$. Refer to Chapter 3 (Section 3.2.1) and Chapter 2 (Figure 2.2). Assume the insulation thickness $\gg$ the air gap and that parallel field conditions are applicable $\varepsilon_r = 2.2$.

Chapter 5
Insulation defects in power-system equipment: Part 2

- Low- and high-voltage motors
- Large generators
- Power, distribution and magnetic voltage transformers

This chapter considers the construction of rotating machines and transformers in sufficient detail to identify a number of possible failure modes. Several examples are included.

5.1 Electrical rotating machines

In the supply systems, a wide range of rotating machines is utilized in addition to the power generators. This is especially the case in the power stations where motors from ratings of less than a kW at 415 V through to those for blower and pump drives of up to the order of 10 MW or more at 11 kV might be installed.

5.1.1 Low-voltage motors

With the wider application of power electronics for adjustable speed drives, there seems to be a need for more careful design of the insulation systems even for the lower-voltage motors. The presence of partial discharges has been detected in random wound stators operating at 440 V supplied from PWM converter circuits [1]. In this area specialized tests are required to prove the turn-to-turn strength. A differential 0.2 μs rise-time surge test [2] as defined for formed wound machines may be applied for assessing the inter-turn wire covering. However, this does not necessarily prove the integrity of a winding subjected to less steep pulses but of much higher repetition rate. The complexity of the conditions as imposed on the insulation are considered in Reference 3. Following failures of motors supplied by PWM control systems, investigations were made of the effects of voltage/frequency variations, pulse rise times, pulse width/duty cycles and overshoot/resonance voltages on the electrical insulation. It is possible for partial discharges to develop during the pulses under adverse conditions. The incorporation of sine-wave output filters may result in improvements but their size and cost can be prohibitive [4]. The work of IEC TC2 WG27 in preparing the new IEC Specification IEC 60034-18-41 [5] is presented by Wheeler in Reference 4.

5.1.2 High-voltage machines

In principle, the electrical insulation systems of the stator windings of the majority of high-voltage machines may be considered to be similar [6]. Insulated conductors, usually of rectangular section, are assembled in the slots of the laminated cylindrical core – a process described as *winding*. As depicted in Figure 5.1(a), the connections between the active parts (straight sections) of the insulated conductors are made in the 'overhang' region, enabling the formation of series/parallel configurations with the appropriate coil pitch (spread) to produce the desired three-phase output voltage. The overhang is supported by various insulation structures, possibly involving large-diameter insulation rings, cords (strings) and additional insulation taping.

5.1.2.1 Motors

For motors in the range of 3 kV to 11 kV and up to tens of MWs ratings, it is possible to preform and insulate the two-sided diamond-shaped coils containing multiple rectangular turns with the appropriate pitch and overhang contour suitable for assembly

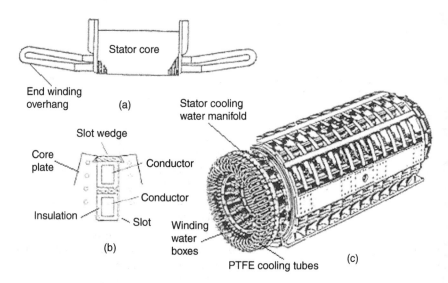

Figure 5.1   Layout of coils/bars and insulation in AC stator winding. (a) Core and end windings (overhang); (b) Turns/bars in slot; (c) Large water-cooled turbine generator stator

(winding) in the core slots. As an example, the conductor dimensions for a three-phase, 11 kV, 3 MW motor might be of the order of 5 × 3 mm cross-section, 4 in parallel with 5 turns/coil and 30 coils per phase having an axial core length of about 1 metre with overhangs of less than 0.5 metre. The core inner diameter might be of the order of 0.75 metre.

5.1.2.2 Generators

The large low-speed (a few hundred rpm), air-cooled, salient-pole, vertical-shaft hydro-generators may have ratings up to 500 MVA at 15 kV and the steam-turbine generators up to 1 600 MVA at, possibly, 27 kV and 1 800 rpm (60 Hz). The latter types form the major power sources for the supply systems of the utilities and more usually have outputs of 500–660 MVA operating at 3 000 rpm (50 Hz) and 22 kV. Each stator bar (two per slot) is manufactured separately with the correct overhang configuration and, after assembly in the core, is connected to its associated turn – probably by means of a flexible link. With internally cooled generators the distilled water is supplied through plastic tubes – typically of PTFE – feeding 'water boxes' at the ends of the bars (Figures 5.1(b) and (c)).

The conductor of an older 500 MVA, 22 kV steam turbine-generator might have an overall cross-section of 66 × 26 mm comprising many strands in parallel each insulated from the others to reduce eddy current losses and hollow for water cooling. The bar lengths might be of the order of 6·5 metre within the core and 0·8 metre for each overhang. Two parallel paths per phase are required with 16 bars in series in each, depending on the number of slots. Full line voltage appears between the line end bars (conductors) which may be physically adjacent in the overhang. Phase voltage to ground is present from the line end bars to the core within the slots. The whole system – core, rotor and windings – operates in dry hydrogen at a pressure of about 6 bar, the gas acting as a coolant and insulant. See Figure 2.2 for the breakdown characteristics of hydrogen.

5.1.2.3 Machine insulation systems

In the older machines, many of which are still in service, the turn insulation consisted of 'thermoplastic' systems. Following many years of research and development, mica paper (incorporating small mica splittings), synthetic thermosetting resins and higher-temperature backing tapes were introduced in the 1960s and continue to be the configurations now used but with improved techniques and material formulations. The three basic methods are briefly outlined in the following.

(a) Shellac/bitumen binding varnishes
Earlier systems comprised mica splittings, cellulosic paper and binders of shellac/bitumen varnish. After wrapping of the insulated conductors and forming of the coil contour – two-sided for smaller machines and half-coil/bars for large generators – the wall insulation was added, the straight sections compressed to the correct thickness for fitting in the core slots, the overhang taping added and the whole consolidated at an appropriate temperature. If not continuously hand-taped, a scarfed

Critical flashover distances

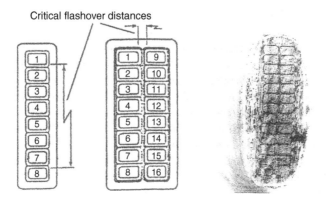

Figure 5.3   Multi-turn motor winding. (a) Single section; (b) Double section; (c) Taping in end-winding of failed 11 kV motor

between sections if wound, as in Figure 5.3(b), due to the non-uniform voltage distribution across the line-end turns. The photograph in Figure 5.3(c) shows part of the overhang section of a VPI coil the impregnation of which was insufficient. A failure occurred in a similar manner to that in Figure 5.3(a), due to switching of a vacuum circuit breaker at the end of a cable. Protective devices were added after this incident.

In large machines the interaction of the distributed capacitance and inductance is important when considering the analysis of partial-discharge measurements. The results relate to such parameters as the non-uniform pulse distribution, the winding pulse transit time and the coupling between phases. These factors also influence the designs of the test systems for monitoring the partial discharges.

5.1.3.3   End-winding insulation

The insulation of the winding overhang can be complex. This is associated with the necessary arrangement and shape of the conductors together with the asymmetrical electromagnetic forces set up by the stator winding currents. In the case of motors many starts per day are often required. Continuous vibration can lead to the spacer blocks becoming loose, resulting in fretting of the conductor insulation. Dust due to this process is evidence of insulation deterioration. Although the electrical stresses to ground are much lower than in the slots, there are locations in many designs where line-to-line conductors are adjacent and insulation-creep surface strengths must be maintained – particularly under poor environmental conditions (e.g. moisture and dust). Failures in these regions have been reported. The presence of creep surfaces can represent a design problem for large hydrogen-cooled generators, which are tested in the factory and at site in air at the $2U_N$ level. It must be accepted that the newer resin-impregnated materials (e.g. glass-fibre systems), as well as the natural types such as impregnated laminated wood, will absorb a limited amount of moisture, which could result in a degradation of electrical properties.

A one-minute test in air on a hydrogen-cooled generator at approximately $2U_N$ (AC) resulted in discharges, initiated at the semiconductor edges, flashing across the end insulation surfaces. This was probably due to localized charged areas on the high-resistivity insulating materials used. These effects are not present in hydrogen at the operating pressure of 6 bar, as the gas breakdown stresses are much higher for this condition.

To minimize surface PDs at the exit point of the coils/bars (Figure 5.2 B), a semiconducting surface is added on machines of about 6 kV and above. If deteriorated or incorrectly applied this can result in high-energy partial discharges including air-corona at operating voltages. This was observed for an 11 kV pump motor where excessive corona was located around the end windings using a handheld ultrasonic detector. The levels were too high to carry out significant PD tests. Such a state may have been causing environmental problems within and external to the unit.

Also in these exit regions it may be necessary to incorporate a scarf joint (Figure 5.2 B) for integrating the slot wall insulation and the overhang external tapes. In another type of 11 kV motor semi-cured tapes had been applied in the formation of the joint in order to assist in winding. During assembly it was suspected that undue force was applied, resulting in a lack of bonding during the final cure. A failure occurred at such a location in service. Subsequent testing of the other phases indicated partial discharge levels of several thousand picocoulombs at below operating voltage. Subsequently, failure occurred at the corresponding joint on increasing the test voltage. The mica paper was found to be poorly bonded.

5.1.4   *CIGRE summary of expected machine insulation degradation*

It is of value to note the summary included in CIGRE General Reports for 1998 [9].

The common opinion is that modern epoxy-mica based insulation systems will not fail for electrical reasons. What will affect the insulation is mechanical malfunction that may in time lead to electrical failure. Such failures include:

- Excessive coil end vibrations
- Loosening of the bars in the slots
- Loosening of the end winding bracing
- Water leakage resulting in delamination of the insulation of water cooled bars
- Poor manufacturing quality for earlier installations (prior to 1970)
- Failure of the semi-conducting layer of the stator bars

Thus, for modern systems, the insulation in itself cannot be expected to be a root cause of future failures. This is in contrast to the experience with many older systems where electrical problems are a common cause of failure.

As the fleet of operating generators includes many old as well as new machines, it can be expected that a lot of electrical as well as mechanically induced failures will occur in the future.

5.1.5   *Future of machine insulation*

The development of improved insulating-tape systems continues, as does the advancement of stator insulation manufacturing techniques for large units – for example VPI processes. A new concept utilizing high-voltage XLPE cable as the stator winding thus allowing operation at much higher voltages is described in Reference 10. The prototype was commissioned in 1998, having a rating of 11 MVA at 45 kV and 600 rpm. The first commercial units included a turbine generator rated at 136 kV, 42 MVA and 3000 rpm and a hydro-generator of 155 kV, 75 MVA and 125 rpm rating. Applying the same principle, high-voltage motors within the rating range 4 MW to 70 MW are available.

Monitoring of the large machine insulation systems is expected to expand as the measurement techniques become more advanced. Some possibilities are discussed in Chapter 9.

5.2   Transformers and reactors

The insulation problems associated with power-system transformers and reactors are especially related to the complexity of the various windings, the methods of cooling, the need for efficient clamping techniques to resist short circuits and transportation forces, the location and mounting of interconnecting leads, the accommodation of bushings and tap changers, and the form and condition of the insulating materials themselves. All these factors influence the forms of the three- dimensional electric fields and the magnitude of stresses in the insulation structures as set up by the steady-state voltages and, critically, the surges due to lightning and switching.

Insulation deterioration in service occurs naturally but the rate can be reduced by good and well-planned maintenance. Although the situation is improving with the introduction of more extensive monitoring, transformers are sometimes treated as the 'sleeping partners' in the system, even though their continuous service is vital to the successful operation – both technologically and economically – and to the versatility of the networks. The latter include DC transmission schemes where converter transformers form the link with the AC generation source (Chapter 1).

5.2.1   Windings

5.2.1.1   General considerations

In principle there are two basic forms of transformer winding: (i) helical and layer types with turns wound axially in single or interconnected multiple layers; and (ii) disc (core-type transformers) or pancake (shell-type transformers) types with the turns wound radially and the discs continuous or coils interconnected if wound separately. There are many winding variations depending on the voltage and current ratings, the tapping range specified, the reactance required between windings (an important parameter in system design), any switching surge tests (see Chapter 1), the likelihood of very fast transients (VFT) associated with SF_6 switchgear and, very importantly, the lightning impulse test (Chapters 1, 6 and 7).

5.2.1.2   Lightning impulse design

The impulse test requirement has a major influence on the design of high-voltage windings as the voltage distribution between sections, layers and turns is not identical to that for the power-frequency condition. This resulted in the use of shielding to reduce the effective shunt capacitance C_g, interleaving [11,12] to increase the effective series capacitance C_s or judicious design with large rating disc windings to increase the radial depth, thereby producing a natural increase in C_s. The effect of these parameters on the initial distribution of the surge voltage – represented as a unit function – may be demonstrated by considering the equivalent capacitance network for a uniform winding (Figure 5.4(a)). This simplification assumes the series inductances of the winding are open circuit at zero time. The final voltages are theoretically proportional to the resistive components and are thus identical to the turns distribution. In Figure 5.4(b) are plotted voltages for different values of α where $\alpha = \sqrt{\frac{C_g}{C_s}}$ in the equation $v_x = V\{\sinh(\alpha x/L)\}/\sinh \alpha$. Note that C_g and C_s are the effective total values and that the winding has an earthed neutral. By achieving a low α value a better balance between the insulation requirements for steady-state and impulse conditions is possible. At some locations special precautions are necessary to allow for the surge

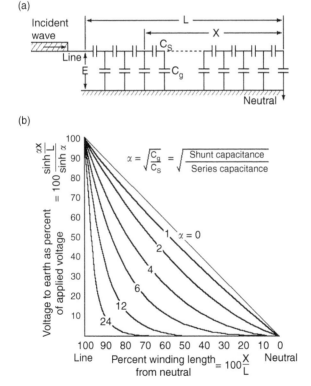

Figure 5.4   Initial distribution of impulse voltage in a uniform winding with earthed neutral. (a) Equivalent capacitance network neglecting inductances and losses; (b) Initial distributions for different values of factor α

indoor applications in the 150–300 kVA, 17 kV/415 V range are now common. Many of the small 5–10 kVA, 11 kV/415 V pole-mounted transformers are insulated with oil-based systems but these will probably be replaced by dry-type designs.

Although not considered here, a number of SF_6 gas-insulated power transformers are in service. A notable example is the substation in the business centre of the city of Sydney (Australia), where three 400 MVA, 330/132 kV units were installed in 2003 [22].

Overall conditions that can contribute to oil-filled transformer insulation failures are (i) the presence of moisture during test caused by inadequate processing and, on site, due to inappropriate maintenance or long-term ageing, and (ii) the trapping of air caused by incorrect vacuum filling in the factory and on site, the latter perhaps after lowering of the oil for inspection purposes or a bushing change. In some dry-type designs, air clearances and termination configurations may be such as to cause corona and consequential erosion of the adjacent insulating materials.

5.2.2.1   Windings

(a) Low-kVA oil-impregnated windings
The high-voltage windings of the low-kVA units (11 kV systems) may consist of resin-covered, small-diameter wire wound in layers with a low-voltage, single-layer winding of enamelled rectangular strip. In some cases the LV winding comprises layers of aluminium foil separated by one or two sheets of presspaper of, perhaps, 0.12 mm thick. Although the volts/turn are very low – perhaps five – a few cases have been reported of inter-turn service failures, probably due to metal particles trapped in the rolls of foil puncturing the paper during operation but not on test. In another case, from a different manufacturer, an 11 kV foil winding of a well-established design failed during impulse test. The inter-foil (inter-turn) insulation was 0.2 mm thickness (several tens of volts/turn) specially prepared to allow resin curing and oil penetration during subsequent vacuum filling. The failure was due to unusual mechanical distortion of the foil during winding. Foil thicknesses seem to be in the range of 0.4 to 0.6 mm, depending on the current rating. A more conventional 11 kV, 7 MVA three-phase winding might have eight layers of copper, resin-coated, rectangular conductor with dimensions of the order of 11 mm × 4 mm × 4 strands in parallel and inter-layer presspaper containing vertical cooling duct(s). The system would be dried and oil-filled under vacuum.

A magnetic-type voltage transformer (MVT) consists of a bushing, a core and windings with high turns ratios. The high-voltage winding conductor is of fine wire, perhaps of the order of 0.5 mm diameter, usually wound in concentric layers. A good impulse-voltage response is achieved because of the high capacitances between layers and by the provision of inner and outer electrostatic shields. For the highest voltages a cascade arrangement is preferred.

Resin- and SF_6-insulated units are in service. Efficient resin impregnation around a large number of turns of fine wire requires very careful processing.

A number of major failures of oil–paper MVTs have been associated with the termination of the ends of the electrostatic shield on the outside of the high-voltage

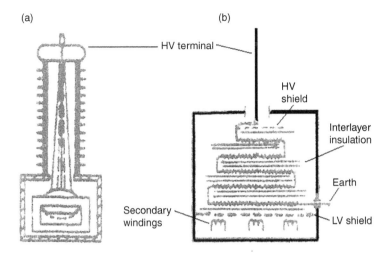

Figure 5.6   Single-stage magnetic voltage transformer (MVT). (a) Layout of dead tank design with layer winding; (b) Equivalent circuit representing many concentric layers

winding. The drying and vacuum oil impregnation of the ductless paper insulation between winding layers must be efficient to prevent voids and subsequent partial discharges. The form of a particular single-stage structure is shown in Figure 5.6. In some cases ducts and layer end insulation may be incorporated

(b) Dry type windings
Small, dry-type windings for special applications are insulated with materials such as polyester film (Class 130, e.g. Mylar and Melinex) and composites and aramid paper (Class 220, e.g. Nomex) and then impregnated or simply dipped in resins followed by curing. The latter procedures can lead to the trapping of air with resultant partial discharges of many thousands of picocoulombs in the inter-winding insulation. A number of such units failed in service due to this cause. PD testing enabled recommendations to be made for improvements in the design and processing methods. For larger ratings of, perhaps, 150 kVA (three-phase) at 22 kV, VPI techniques have been developed. This allows complete encasement of each phase of the HV windings in the resin using a mould giving a smooth outer finish with appropriate connections. Each of the windings is lowered over the particular LV winding on the core with a corona-free inter-winding air gap. PD levels of less than 10pC are achievable although connecting leads and mountings have produced much higher values – as located by ultrasonic detectors.

(c) Higher-rated oil-impregnated windings
Low-voltage windings (11–22 kV) for oil/paper/pressboard-insulated transformers of higher ratings are usually of the helical form in one layer with many strands in parallel. Problems have occurred where the enamelling on the conductors was

relied upon to provide insulation between strands but after several years of service abrasive mechanical movement between the conductors produced short circuits with subsequent circulating currents and overheating. This required complete rewinding with paper-insulated conductors at considerable cost. Such faults might be detected by monitoring of gases and furans produced by overheating of the enamel, pressboard spacers and adjacent oil.

The larger layer-type high-voltage windings in core type transformers for 220 kV and above (Figure 5.5(c)) might consist of 5–7 layers of paper-insulated, continuously transposed cable. Extra layers are included for taps and the low-voltage section of an autotransformer. The turns are wound on wraps (sheets) of paper up to a thickness of possibly 12 mm and cooled by oil ducts maintained with axially located pressboard spacers of 6 to 12 mm thickness, the relative dimensions depending on the electrical and thermal design requirements for the particular unit. With an earthed neutral, the layers and insulation are graded as in Figure 5.5(c). The inter-layer paper is flanged (petalled) at right angles to form the end insulation including a continuation of the vertical oil ducts. In one arrangement the layer ends are brought out through the horizontal ducts and connected in series – 'top to bottom' – by the provision outside the winding shield of accurately located insulated leads of specified dimensions [13]. The inner and outer shields are necessary to ensure a good impulse distribution for the simple arrangement shown. The inner shield may not be required under some conditions. Insulation faults in these windings include deterioration of the winding-to-shield connections, blocking of ducts due to bulging of paper at the end of the layers and lead connection problems causing local overheating. Gas analysis is the major monitoring technique used, although some faults have been located ultrasonically. Tracking due to electrostatic charging in the lower inlet and cooling ducts was reported by one manufacturer. Tests on models and in-service transformers were carried out [15]. The problem was located by partial-discharge ultrasonic methods and confirmed on unwinding. It is considered that the layer-type winding has been successful with units operating for 40 years or more. However, it appears that the international mergers of manufacturers and the associated economic evaluations will see the demise of the design in favour of the more versatile disc winding.

Shell-type transformer coils (Figure 5.5(d)) have good surge response and are well supported mechanically [14]. The ducts between 'pancakes' are relatively thin and the cooling oil-flow rates tend to be high. Electrostatic charging effects were believed to have resulted in a number of service failures in different types of transformers in which local oil velocities seemed to be excessive. Extensive investigations in the USA and Japan and by CIGRE Study Committees 12 and 15 were carried out [16,17]. Some monitoring is possible by low-voltage DC measurements and ultrasonic detection. The advantages of shell-type transformers are being applied by some manufacturers.

The majority of high-voltage windings from tens of MVA (66 kV) up to the largest ratings and highest voltages consist of discs of paper-insulated conductors wound continuously or in pairs (coils). Horizontal and vertical ducts are formed by pressboard spacers. With directed flow units, ducts on the inside and outside diameters of the winding are formed by vertical barriers (cylinders) and insulation washers provided at critical locations in the stack in order to direct the oil flow horizontally between

(a) (b)

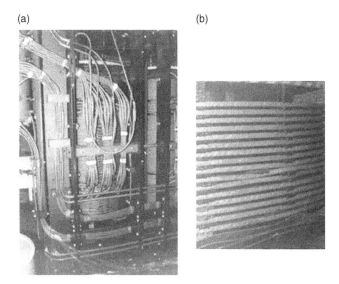

Figure 5.7   Example of the surface of a medium-size disc winding (10 MVA, 66 kV).
(a) View of leads from tapping windings located within the 66 kV stacks.
(b) Displacement of 66 kV winding turns due to service fault

discs. The oil is pumped into the bottom of the winding from the external air or water coolers.

The failure rates of windings are low, as reported in CIGRE surveys and papers. Problems with the older medium-size units are probably associated with lightning strikes where the protection was inadequate or the design not appropriate. Windings may fail also due to tap changer maloperation (Figure 5.7). Occasionally, unusual faults occur as in a 66 kV winding where a nut had lodged between a disc and a vertical barrier. This probably occurred during an inspection three years previously, but was not detected by the overvoltage tests applied at that time. It was deduced that the nut was dislodged into a more critical position during service and partial discharges were set up, which eventually led to a winding failure. Such a condition may have been detectable by gas analyses or, possibly, by PD measurements on site. It would have been very difficult, or impossible, to detect by ultrasonic methods, as the fault was within the winding. The relocation of aged, moderate-size or large, transformers can produce internal movement and consequent insulation damage. This problem has been reported for new transformers, where insufficient allowance had been made for the forces produced during transport – by ship, rail and road.

The desirability of continuous online monitoring, especially for large generator and transmission transformers, is indicated by a number of failure types whose causes are difficult to determine with certainty. These include (i) manufacturing faults related to the winding insulation, which develop into a dangerous condition only after some time in service, e.g. mechanical failures of conductor joints; (ii) electrostatic charging

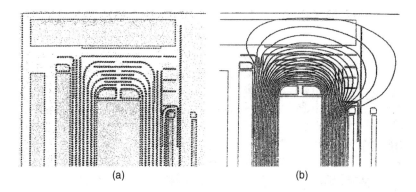

(a)   (b)

Figure 5.10   Typical insulation layout and equipotential field of an end-connected high-voltage power transformer [diagrams by Weidmann Transformer Systems AG, Rapperswil]

paper wraps have probably exhibited similar problems. By the inclusion of thin ducts (see Section 3.2.1) it is possible to achieve high average operating stresses if this is required in the proposed design. Other faults detected within inter-winding insulation include a broken connection between the inner axial shield and neutral.

A potential fault condition in a newer transformer was the presence of contamination by fine metal particles produced by a faulty cooler. This particular unit had not failed but on disassembling was found to contain pressboard and paper surfaces covered with the particulate. On-site PD tests showed no apparently significant values and, again, after a factory impulse test during which failure occurred well below the expected withstand level. Physical examination detected a fine track mark along the surface of the barrier forming the duct next to the inside diameter of the high-voltage winding (centre entry). In this case conventional electrical monitoring was unable to predict a possible service failure due to the particles or a track path created by a low-energy surge flashover [19].

Metal particles of millimetre dimensions within a structure have been detected during PD tests in the factory. Such a particle will produce PDs only if it is located and trapped in a position where its shape enhances the electric stress. PD magnitudes of the order of hundreds of pCs might be expected [19]. A number of failures in service as well as during overvoltage tests are reported in the CIGRE review [20]. These problems are more likely to be associated with long oil distances, e.g. a bushing termination to turret or the outside of a winding to ground, rather than within the coils themselves.

5.2.2.3   End insulation
With end-connected windings (Figures 5.5(e) and 5.10(a)), careful layout of the insulation is necessary to prevent flashover along surfaces to ground (including the earthed core) and to adjacent windings such as outer taps and to ensure good oil flow for cooling of the windings. The use of a radial shield connected to the line end disc

(Figure 5.10(a)) [21] and the contoured cylindrical pressboard barriers following the equipotential lines (Figure 5.10(b)) provide an efficient insulation system. In Figure 5.10(b) the equipotentials represent 5 per cent intervals. The corresponding electric stresses may be calculated by means of an associated field computer program. Ideally, the spacings between barriers may be staggered in order to prevent partial discharges in the particular oil gap (Figure 2.3) during testing. This is especially important adjacent to the maximum-stress areas at the corners of the high-voltage and low-voltage windings and the tap windings. In Figure 5.10(b) shields at the corners of the LV and tap windings are at, or near, earth potential. At the position of connection of the HV lead the configuration becomes more complex. Preformed barriers and components for lead insulation are available from specialist manufacturers. The figure's plot is for a separate source test. During an induced or impulse test conditions in the end insulation do not change significantly except, perhaps, at the lower corners of the radial shield.

In a particular case omission of horizontal pressboard washers above the cooling gap through the radial shield (Figure 5.10(a)) resulted in an impulse test failure along a tortuous pressboard path to earth.

A problem that has arisen is the production of partial discharges at the edges of the metal (earthed) winding clamping studs where these were not of correct dimensions and insulated. An unsatisfactory arrangement can result in PDs of many thousands of pC leading to flashover to an adjacent high-voltage lead or similar during factory tests (see Figure 3.6). Discharges at the surface of earthed clamping studs have been detected by DGA on-site and located using ultrasonic methods. During maintenance, the upper insulating board (laminated wood or pressboard) may be exposed when lowering the oil, thus allowing ingress of moisture if precautions are not taken. Failures due to dielectric losses in such boards are possible as considered in Chapter 2.

5.2.2.4 Insulation to tank and core

It is usual in the larger transformers to provide pressboard barriers in order to break up the long oil paths between the windings and tank, between phases (three-phase) and to the outer core leg (5-limb designs), especially where particles might be present [20]. The distances involved are possibly in the range of 50–250 mm depending on the test voltages and the chosen safety factor. In older transformers tracking on barriers has been found although no 'direct feed' point was involved. The effect is probably due to moisture development or ingress and would not be detectable by existing electrical monitoring [19]. Some change in DGA sample values may be measurable.

5.2.2.5 Bushing internal terminations and leads

Bushing designs have been discussed in Section 4.3. The method of connection of a conventional bushing within a power transformer at high voltages is indicated in Figure 5.5(e) and the associated field plots for both conventional and re-entrant types mounted in a turret in Figure 4.3. Plots are also presented in Reference 13, including the equipotential distribution for a re-entrant bushing without a high-voltage lead in place. Bushings installed in transformers are shown in Figure 5.11.

(a)   (b)

Figure 5.11   Bushings installed in high-voltage power transformers. (a) 100 MVA, 11/230 kV generator transformer [photograph by Wilson Transformer Company]; (b) 133 MVA, 330/√3/138/√3 kV auto transformer [by permission of Transgrid NSW]

Flashovers have occurred between bushings and turrets due to contaminated oil and particles and where a local barrier has been exposed for too long during installation. In a number of cases the stress distributor (Figures 5.5(e) and 4.3(a)) has become loose, resulting in discharging (sparking) and gas production, which might have triggered an axial flashover. Such discharging has been detected by DGA measurements and successfully located by ultrasonic techniques. Problems in re-entrant bushings can arise due to incorrect location of the paper-insulated lead within the porcelain (Figure 4.3(b)). For example, air might be trapped within the paper–porcelain gap in some configurations when topping up the oil on site.

The design and insulation of leads from bushing ends to the windings require special care to ensure safe stresses at the paper surfaces. By use of prefabricated pressboard systems, concentric cylinders and angles can be built in as shown in principle in Figure 5.5(f, 2/3). A critical region in such structures is where a scarfed joint is required at the exit point. Creep along an incorrectly made scarf can lead to a major flashover with pre-discharging of thousands of pCs during AC tests. The discharges may be detected by PD measurements and located by ultrasonic methods. Failures to the top of unshielded tap windings without barrier protection have also occurred on test, the condition being exacerbated by the '*radial*' stressing produced by the adjacent high-voltage winding (Figure 5.5(e) and (f, 2/3)).

The location, the thickness and the diameter of insulation around high- and low-voltage leads are determined by the local field distortion as well as the oil and creep paths. Field analyses are necessary to ensure the stresses are acceptable.

5.2.2.6 Tap changers

Although most tap-changer failures can be traced to mechanical maloperation, some are related to tracking along the insulation surfaces or a direct oil flashover perhaps associated with contamination created by incorrect wearing of the metal contacts.

In a particular 66 kV transformer, distortion of the tap windings in the centre of the high-voltage winding (Figure 5.5(b)) was due to a flashover along the inner cylinder of the 20-year-old tap changer. Some damage was also present in the main windings, thus requiring a rewind of the complete leg.

As conditions within a tap changer are quite onerous, monitoring of the gas levels and regular maintenance are essential. New techniques are being developed to improve surveillance. Care is necessary when interpreting transformer DGA results if the tap changer is in the main tank or there is a possibility of leakage from the tap changer/diverter enclosures.

5.2.2.7 Core and magnetic shields

Overheating due to circulating currents in cores and laminated magnetic shields caused by low-level interlaminar insulation degradation has been detected by DGA sampling. The temperatures produced can result in breakdown of the oil. The locations have been identified using ultrasonics and units subsequently kept in service until remedial action could be taken. If allowed to build up, the deposits at the fault are difficult to remove as experienced in an old 275 kV, 120 MVA transmission transformer. With magnetic-shield faults, continued operation may be possible at reduced load current.

5.2.2.8 Major transformer failures

Service failures of oil-filled transformers can result in complete destruction of the system by fire and irreparable damage to core and tank as well as expensive outages. For example, see Figure 5.12. If the supply-system protection operates quickly, damage may not involve the whole unit, although carbon contamination can be extensive, especially if pumped oil coolers are installed. It is anticipated that development of continuous monitoring systems will help eliminate such catastrophic failures.

5.2.2.9 Power transformer summary

The situations described are typical of some of the insulation faults that might develop in power transformers. Although a particular failure can be very significant it should be appreciated that the problems involve a small percentage of the total number of units in service. However, it is probable that the percentage will increase as the assets age, full-load (or more) currents are demanded for longer periods and higher electric stresses tend to be applied in new designs to meet economic demands. It is essential, therefore, that testing, insulation-condition assessment and monitoring continue to be improved, especially for the generator transformers and the medium- to large-size transmission transformers. It would be expected that if large SF_6-insulated units [22] become more common some of the existing operational problems will be reduced.

22. Ebb, G.M., and Spence, G.S., 'Gas insulated transformers for Haymarket Substation', *Proceedings of the IEEE/PES Conference on Transmission and Distribution*, Japan, October 2002, vol. 1, pp. 511–16

23. CIGRE WG D1.11 Allan, D. (convener), 'Service aged insulation guidelines on managing the ageing process', *Electra*, June 2003;(208):69–72 (Technical Brochure 228).

5.5   Problems

1.   (i)   Differentiate between the conventional insulation and cooling systems of a large steam turbine generator and that of a hydro-generator.
    (ii)   How are the surface discharges controlled at the exit of the stator bars from the core slots?
    (iii)   Describe techniques suitable for the monitoring of possible partial discharges in the two types of machine (refer to Chapter 9).

2.   In an air-cooled, medium-size rotating machine the outer earthed conducting layer on a stator bar has been eroded by vibration within a core slot. Determine whether air discharges would be expected for an insulation thickness of 3 mm, a spacing to the core laminations of 0.5 mm and a voltage of $11/\sqrt{3}$ kV. If similar dimensions were used in a hydrogen-cooled machine at a pressure of 5 bar, at what voltage might PDs be initiated for a similar condition? Refer to Figure 2.2. Assume a pressboard permittivity of 3.5.

3.   (i)   Describe the major differences between the insulation systems of large core- and shell-type transformers.
    (ii)   The spacing (duct) between adjacent discs (7 per cent of winding turns) at the line end of a uniformly wound core type transformer is 8 mm of oil and the two-side conductor insulation thickness is 1.5 mm. Calculate the theoretical lightning impulse breakdown strengths of two windings of similar configurations with α values of 4 and 1. Assume a uniform field (a big approximation) and that failure would be expected to occur in the oil at the line end to the adjacent disc. The permittivities of the oil and conductor paper are 2.2 and 3.2 respectively. Determine for which voltage class each winding would be suitable assuming a minimum test safety factor of 1.3. Refer to Equation 3.1, Figure 2.3 (c), Figure 5.4 and Table 1.2.

4.   From a transformer DGA results it was found that the oil hydrogen level was 1 200 ppm and the total combustible gas 3 000 ppm. This indicated a source of PDs within the unit. Describe methods for detecting and locating the site as well as confirming that the disturbance was probably a partial discharge. Refer to Chapters 6, 9 and 10.

Chapter 6
Basic methods for insulation assessment

- High-voltage test supplies
- Electrical non-destructive test methods
- Physical and chemical assessment techniques

In this chapter an overview is given of the more important methods employed in the supply industry for assessing the condition of insulation in power-system equipment – before leaving the factory, on commissioning, during service and when undergoing major maintenance or repair. The particular requirements for equipment of different insulation structures are described in Chapter 7, including reference to Standards where appropriate. More advanced recently developed methods are presented in Chapters 8–10.

The earliest methods for determining whether or not an insulating material was suitable for a particular usage included the application of steady-state voltages for, perhaps, one minute at twice or more the operating stress. Later, this was followed by the development of surge voltage tests to simulate lightning and system-switching effects. Much research was necessary in the design and construction of high-voltage test equipment and associated test/laboratory areas.

In recent years there has been a tendency to close high-voltage laboratories in those countries with developed power systems. However, this may be a short-sighted policy, since it is essential to provide local reference measurement systems and capabilities for checking and proving ageing equipment, as well as new items of plant in which unproven insulating materials may be included.

In the following, the various methods of assessment are grouped according to whether the application is:

(i)   an overvoltage withstand test;
(ii) a non-destructive electrical measurement; or
(iii) a physical/chemical determination that may require access to a sample of the insulating material.

6.1   Generation and measurement of test high voltages

The form of the source and method of voltage measurement are briefly described for each type of test voltage. More detailed treatments will be found in the books by, for example, Kuffel and Zaengl [1], Schwab [2], Kind [3] and Hylton-Cavallius [4].

6.1.1   Power-frequency voltages

Power-frequency voltages for insulation withstand testing are produced by specialised transformers with outputs within the range of tens of kV to a few hundred kV in a single unit and up to a million kV or more in cascade arrangements. The voltages required for routine testing are tabulated in Chapter 1. The current ratings vary according to the capacitances of the particular equipment to be tested and the kVA ratings from 10 kVA (e.g. 0.1 A at 100 kV) up to 500–1 000 kVA at the higher voltages. The ratings quoted are maxima and the current values must not be exceeded – for example, a 100 kV, 10 kVA transformer can only operate at 5 kVA (0.1 A) at 50 kV unless otherwise specified. Such transformers are often designed for short time ratings at full-load current and are not overvoltage-tested at the same margins as power transformers. Usually they are not subjected to lightning and switching type surges. Modern units are supplied with a low partial-discharge guarantee, typically less than a few pCs at full voltage. In addition to units produced by test equipment specialists, a number of power and distribution transformer manufacturers have built their own testing transformers.

6.1.1.1   Single-unit testing transformers

The simplest testing transformer consists of a closed core with a low-voltage winding (240/120 V), over which is wound a graded layer winding of many turns, the major insulation comprising oil-impregnated paper. The high-voltage winding is provided with an outer shield (see Figure 5.5(c)) in order to ensure a 'uniform' voltage distribution following a flashover at the HV terminal due to breakdown of a test object or a measurement sphere gap. The core and coils are mounted in a steel tank with a high-voltage bushing (Figure 6.1) or in many applications, for indoor use, in an insulating cylinder with a spinning as the HV terminal mounted on the top of the cylinder. The unit in Figure 6.2 has a diameter of 1 metre. In one case a centre-entry design with an 'open' core and outer insulating cylinder was successfully used for separate source testing up to voltages of 70 kV. Single units for production of a million volts were built in the 1930s but some eventually failed, probably due to partial discharges. The more common single-unit designs are probably in the range of 200 to 600 and a few up to 750 kV [4].

6.1.1.2   Multiple-unit (cascade) testing transformers

Production of voltages above 350–500 kV is probably achieved more reliably by cascading transformers of lower voltages, as indicated in Figure 6.3. The individual units are similar to single transformers but have an exciting winding supplying an upper transformer, which is insulated from earth for the output of the lower unit.

*Figure 6.1   200 kV testing trans-
former – metal tank*

*Figure 6.2   250 kV testing transformer
and capacitor divider – insulation cylinders as shells*

An insulating shell with the individual units stacked above one another to form a composite transformer may be used – for example the 2.4 MV unit at IREQ [1,4] or, more commonly, stand-off insulators supporting the tanks of the upper units. It is important to allow for the different operating currents in the units – the lower unit supplying the power for the other transformers – and also for the non-uniform voltage distribution created across the various windings on flashover at the output terminal. Compensating reactors are sometimes included in parallel with the windings when testing large capacitances, such as cables, where currents of several amperes may be required. Analyses of the various conditions are presented in References 1 and 4.

Although cascaded transformers may occupy considerable floor space, this is offset by the versatility of having available two or three individual units and ease in handling when tests are required at lower voltages. In some applications lower-voltage units might be utilized (e.g. 100 kV each) in order to form a versatile source for site applications.

6.1.1.3   Operation of testing transformers

The simplest supply for a testing transformer consists of a variable voltage 50/60 Hz source such as a 240 V 'variac' transformer for small units up to 5 kVA. An alternative

It is important to note that in test circuits involving testing transformers complete reliance should never be given to a voltage reading on the low-voltage winding as a measure of the high-voltage output. The apparent turns ratio may not be correct or constant under different load conditions due to resonance effects. A high-voltage device must be included and, preferably, a second system – e.g. a sphere gap – used as a cross reference. Partial failures in the measurement system may not be immediately apparent perhaps resulting in an excessive voltage being applied to the test object or an unexpected air flashover.

6.1.1.4   Series resonant circuits

The application of tunable series resonant circuits for high-voltage testing of large-capacitive objects such as cables, power capacitors, generators and high-voltage shunt reactors is well established. The basic test circuits are shown in Figure 6.5(a) and the two possibilities for testing capacitive-type equipment in Figure 6.5(b) and (c). The natural frequencies of the circuits are given by $f_n = 1/(2\,\pi\,\sqrt{LC})$, where L is the HV inductance and C the total capacitance.

The supply circuits are only required to provide the resistive losses in the high-voltage components. A major advantage of the configurations is that they are tuned specifically to the test frequency, thereby avoiding the possibility of unwanted resonances, which can produce overvoltages and even failures in transformer systems – as well as waveform distortion. Additionally, if breakdown of the test object occurs the high-voltage collapses, thus minimizing the damage and reducing the risk of explosion.

The disadvantage of the techniques for general high-voltage testing is the lack of versatility compared with a transformer circuit. A given resonant circuit is limited by the range of load values for which it is tuneable.

Commercial series resonant test sets particularly aimed at cable testing have ratings of up to tens of MVA and voltages of hundreds of kilovolts incorporating several reactors in series. In the usual series circuit (ACRL) the reactor has a split core (Figure 6.5(b)) which may be varied mechanically to give an overall change in inductance of 10 to 20 [1]. Q values of the order of 30 or higher are possible, requiring an input of a few percent of the main circuit rating and a transformer voltage of only $1/Q$ to give the desired output. Individual units of 350 kV have been stacked in vertical assemblies to produce up to 2.1 MV (six units) [1].

In one location a tuning capacitor with a voltage rating of 1000 kV was used for testing high-voltage shunt reactors, e.g. a 33 MVAR, $500/\sqrt{3}$ kV unit.

A more recent development has been the introduction of variable-input frequency systems with fixed values of series inductance (Figure 6.5 (c)). The required test frequencies are within the 20–300 Hz range as specified and discussed in IEC 62067 [S6/3] for tests on extruded cables above 150 kV. It would be expected that the weight/power ratio for such systems would be less than the equivalent variable inductance configuration [8]. The ratios vary from 1 to 8 kg/kVA for ACRL circuits and ≥ 0.6 kg/kVA for ACFL circuits. Q factors for these systems can be > 100 depending on the test frequency and the losses in the series reactor/transformer/load capacitors

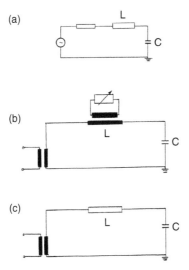

Figure 6.5   *Principle of series resonant test circuits: (a) basic circuit; (b) tuneable reactor circuit (ACRL) – fixed frequency; (c) tuneable frequency circuit (ACFL) – fixed series inductance*

at the particular f_n value [9]. For PD testing applications the IGBT switching pulses in the variable frequency source can be suppressed.

Parallel resonant circuits may be utilized for cable and generator testing on site in order to reduce the supply current. However, this arrangement requires a transformer of the full test voltage – but not full current – as well as a tuneable high-voltage reactor. Tuned circuits are not suitable for tests during which excessive losses can develop, as in pollution testing [4]. An appreciable voltage drop may be produced by the increase in resistive current near flashover.

6.1.2   High-frequency voltages

High-frequency voltages are sometimes used for testing low-capacitance samples, in particular overhead line insulators (see Section 7.1.1). Such voltages may be produced by a Tesla coil circuit in which a charged capacitor is discharged by an air gap through the primary of an air-cored step-up transformer. The secondary is tuned by a second capacitor to respond to the primary current oscillations. Thus a series of high-frequency, high-voltage oscillations are produced [10]. The HV winding comprises many turns of thin wire. Frequencies in the hundreds of kilohertz range and voltages up to a million are achievable with low-capacitive loads. The application of the circuit is very limited because of the oscillatory pulse waveforms and the interference created by the repetitive firing of the air gap. The system is an impressive tool for demonstration of the generation of very high voltages in the laboratory – continuous sparking can be maintained!

6.1.3   Very-low-frequency voltages (VLF)

Following various investigations attempting to relate the strength and PD character-istics of cable and machine insulation at 50/60 Hz to those at low frequency, a number of test supplies have been developed operating at 0.1 Hz.

VLF testing has been used in the world for many applications requiring AC testing of high-capacitance loads. Since HV DC tests can cause space charge and damage to XLPE insulation, VLF has become a widely adopted testing method for XLPE power cables. It is also an effective tool for burning down cable defects in the insulation as well as accessories.

It seems that the low-frequency PD characteristics may not always be similar to the system operating frequencies. This could be due to the large difference in rate of rise of the applied voltages, the higher stress required for inception and the longer time of application to record significant voltage reversals. VLF testing has proved to be of value for detecting electrical tree growth in XLPE insulation.

Different principles are invoked for production of VLF voltages including gener-ation of cosine-square waves followed by polarity reversal giving a cosine function or, more recently, VLF sine wave voltages generated by electronic inverters.

6.1.4   Direct voltages

Direct-voltage testing circuits are necessary for checking the integrity of AC paper-insulated cables and generator insulation in addition to that for DC transmission schemes. The simplest arrangement for outputs in the 100–200 kV range and tens of milliamps is a half-wave rectifier (Figure 6.6(a)) and, commonly, a voltage doubler (Figure 6.6(b)) in which the output is twice the peak of the AC voltage. If required, a smoother output can be obtained by incorporating a centre-tapped transformer. For higher voltages, several stages are included in multiple columns comprising the capacitors, interconnected by solid-state rectifiers. A two-stage system with asso-ciated circuitry is depicted in Figure 6.6(c). The output is four times the AC peak voltage.

Cascaded systems have been built for testing and research up to the voltages required for the various DC transmission schemes. Analysis of these systems is com-plex and it is difficult to predict from theoretical considerations the precise output voltage and ripple content [1,4].

With large-capacitive loads a smoothing (reservoir) capacitor is probably not required, as the test object becomes the dominant component. A series-protective resistor is provided in case of external flashover or failure of the test object. The voltage output is measured by means of a high-value stable resistor connected to a meter calibrated in kilovolts. If the configuration is enclosed in an insulating cylinder of high-resistivity material, care is necessary to avoid static charging effects on the surfaces. In all direct-voltage testing the method of earthing the test object after dis-connecting the supply is very critical. For large capacitances it is probably necessary to discharge the test object through a resistor of sufficient thermal capacity to avoid injury to personnel and to prevent damage due to a sudden current surge.

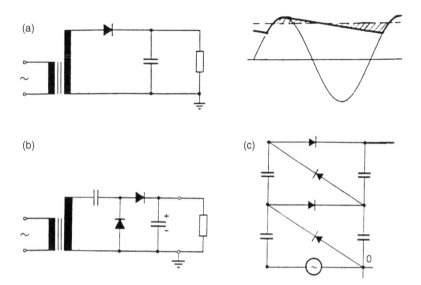

Figure 6.6   Direct voltage testing circuit: (a) half-wave rectifier and waveform; (b) voltage doubler circuit; (c) voltage quadrupler circuit

Portable DC test sets are available for low-current outputs at up to 100 kV or more with a low ripple and high stability. Such units incorporate high-frequency (e.g. 80 kHz) supplies through a small transformer.

6.1.5   Hybrid test circuits

As a possible alternative to resonant test sets and to enable a form of partial discharge measurements to be made on large capacitances, e.g. cables, hybrid circuits have been developed utilizing direct-voltage charging and fast switching to simulate power frequency wave shapes in the insulation system under test.

One method involves the slow DC charging (few seconds) of the test object through a series inductance and when at the required test level shorting the source with a high-speed switch. Due to the low loss of the reactor and cable a number of oscillations can be produced without appreciable decay at frequencies within the range 50 Hz to 1 kHz, depending on the circuit L and C values [11].

A similar system also increases the direct voltage slowly (within 10 seconds) to $2U_0$ and suddenly discharges the test object to produce the required oscillations of about 10 ms per half-period [12].

In both techniques the test voltage is applied several times to obtain significant partial-discharge results.

6.1.6   Lightning impulse voltages

The magnitudes and wave shapes of the transient voltages required for testing power-system equipment are specified in IEC 60071, as described in Chapter 1. The basic

of the first-stage capacitor be earthed directly (not through a resistor), thus giving the same voltage polarities and, incidentally, reducing the generator height (Figure 6.8). Such a reduction may be of significant advantage in a small test area, where only three or four stages might be installed. Very high voltages of up to several million are available in a number of factories and laboratories worldwide [4]. This might be achieved, for example, by 28 stages with a charging voltage of ± 100 kV and capacitors of the order of tens of microfarads to give a maximum output of 2.4 MV at an efficiency of approximately 80 per cent. The capacitors may be in porcelain shells separated by similar spacers and stacked in columns or in metal cans mounted within the insulating framework. One scheme for distributing the front resistors and thereby reducing the voltage stresses across them is given in Figure 6.8. In addition, the tail resistors are more easily supported within the generator structure than externally. The use of distributed resistors is restrictive if frequent wave-shape changes are necessary, but this disadvantage is usually acceptable because of insulation requirements, clearance reasons and the limitation of the effects of internal oscillations. External components can be added if necessary.

In some applications the circuit is allowed to oscillate to produce higher voltages and/or a wave with specified oscillations. A series high-voltage inductor is connected in the output to create an oscillating lightning impulse (OLI) [1].

The circuit in Figure 6.8 includes the DC supply as described in Subsection 6.1.4. This must be suitable for polarity reversal by means of a built-in switching system. It is essential that an automatic earthing switch be provided at the supply terminal – direct voltages can lurk where not expected! The trigger electrode is fired by an external pulse, which also allows synchronism with the recording devices. The latter should comply with the requirements of IEC Publication 61083-1 [S6/4]. CROs and/or digital recorders monitor the output of the HV divider and other transducers if in use. The divider may be a resistive type with an HV arm of perhaps 10–20 kilohms. The lower arm must be correctly matched with the coaxial cable and input impedance of the recording instrument. Capacitance dividers of 100–1 000 pF are often used but allowance is necessary for possible distortion of the incoming signal due to the LV capacitance/coaxial cable surge impedance [1,2,4]. Many investigations have been carried out regarding the performance of dividers, especially with respect to their response to the high-frequency components of the impulse waves. No divider is a pure resistance or capacitance. Various compensating techniques including shielding and additional damping are used at the very high voltages to obtain the required accuracies. At UHV of several megavolts the tall columns and long connections may introduce additional errors in the measurements. Combined resistive and capacitive dividers are utilized under some conditions in order to obtain acceptable responses [1,4].

A sphere gap can be used for measurement or for 'chopping' the output wave at a predetermined time. Synchronization between the generator and main gap triggering is necessary – perhaps from less than 0.5 μs (front of wave testing) to tens of μs. Chopping times repeatable to within 0.1 μs or less may be required. For voltages above about 500 kV multiple chopping gaps are available giving more reliable firing than one large gap.

A major problem with impulse-voltage testing systems is the possibility of inter-ference current pulses being induced, or flowing, in the earth circuits and measurement cables. The configuration indicated in Figure 6.8 assists in minimizing these effects. The latter are discussed in IEC 60060-2 [S6/2].

The objective of such test layouts is to limit the main discharge current of, perhaps, thousands of amps to a very low impedance path and to arrange the divider so its earth connections do not directly involve the high-current path. This configuration may not be successful in particular situations. Additional copper mats or thin copper foils of 10 cm width for earth connections – even with an under-floor system – may be helpful. Double screening of cables together with power isolation and good shielding of the control desk instrumentation is often essential. Experience related to the design and operation of a particular ultra-high-voltage laboratory is reviewed in Reference 4.

6.1.7 Switching surge voltages

For testing of the majority of purely 'capacitive'-type equipment with standard switch-ing surges as specified in Table 1.2 (Chapter 1), a conventional impulse generator may be used, with appropriate modification of the resistive components of the RC circuit. The slow front of $200\,\mu$s requires an increased value of the front resistor and/or an additional capacitance across the test object. Care must be taken to ensure that the energy ratings of the chosen resistors are sufficient to avoid failure by thermal overloading, which might result in flashover along their length.

An alternative method [S6/2] of producing a switching surge is to induce the required voltage in the high-voltage winding of a testing transformer, or a trans-former to be tested, by application of an impulse to the low-voltage winding. For slow-impulse rise times the transferred voltage in a transformer is approximately that of the turns ratio.

A warning in the Standard is given that non-disruptive discharges can occur in the test object during switching surge applications and may result in appreciable wave-shape distortion. The effect can be very significant in pollution tests on external insulation at high voltages. If the source is of high impedance the desired disruptive discharge may not develop.

6.1.8 High-voltage equipment for on-site testing

On-site high-voltage tests are applied during the commissioning stage to ensure equip-ment is undamaged and correctly assembled when new and after locally completed repairs or invasive inspections.

Such HV tests are also required for various diagnostic procedures, in particular partial-discharge and dielectric-dissipation-factor measurements. Some of the site diagnostic requirements for individual items of equipment are discussed in Chapters 8 and 9.

All the sources considered in Sections 6.1.1–6.1.7 may be required on site. However, the normal laboratory/factory testing equipment is usually unsuitable for transportation, and special designs of portable systems are necessary. These may take

Figure 6.9   Example of high-voltage test vehicle (250 kV testing transformer and ancillary equipment) [photograph by Enerserve (Energy Australia)]

the form of custom-built vehicles containing, for example, HV testing transformers and standard capacitors (Figure 6.9), resonant test sets, generator and LV transformer for induced-voltage testing of power transformers, direct-voltage supplies for cable testing (excluding tests on extruded cables) and smaller units for separate source tests (50 Hz, VLF and hybrid supplies) for the lower-voltage systems of 11/33 kV.

It is noted that VLF techniques are used on site for withstand voltage testing of oil–paper and extruded cables and for rotating machines, thus reducing the high-value charging currents required by conventional non-resonant power sources. Voltage outputs are of the order of 50–120 kV, although systems of higher value have been described [13]. The choice of frequencies for site tests is considered in Reference 9.

Impulse generators of modular design are required for testing GIS and some transformers on site. It is possible almost to double the generator output by providing a series inductance between the generator and load and allowing damped oscillations to develop – both for lightning (OLI) and switching (OSI) surges. Frequency ranges for OLI and OSI voltages are suggested in Reference 8. In this report of CIGRE TF 33.03.04, recommendations are made as to how the Standard requirements of IEC Publications 60060-1 and 60060-2, 1994 [S6/2], might be modified to allow for on-site conditions. A number of references of developments are quoted in the article. The Standard specifying definitions and on-site testing requirements was issued in 2006 [S6/2]. The details in Table 6.1 are extracted from Reference 8 and are a guide as to the types of on-site tests that may be specified for different items of equipment. Clearly, an important factor in the design of on-site testing components is the need for greater mechanical strength than their counterparts in the laboratory and also increased awareness regarding safety aspects when setting up temporary test areas.

Table 6.1   Voltage types and sources used for site testing [8] [reproduced with permission of CIGRE]

equipment to be tested on site		cables			GIS GIt.	instrument transformers	power transformers	rotating machines	arresters
		oil-paper cables	extruded cables						
			m. v.	h. v.					
direct voltage (DC)		W	W[1]					W[1]	
very low frequency voltage (VLF)		W	W, DM					W	
alternating voltage (AC)	by ACTC				W, PD	W, PD	W, PD, DM	W, PD, DM	W
	by ACRL	W	W, PD, DM	W, PD	W, PD	W, PD	W, PD, DM	W, PD, DM	W
	by ACRF	W	W, PD, DM	W, PD	W, PD	W, PD	W, PD, DM	W, PD, DM	W
lightning impulse	aperiodic (LI)				W	W			
	oscillating (OLI)				W				
switching impulse	aperiodic (SI)				W				
	oscillating (OSI)		PD[2]		W			W	

W	–	withstand voltage test
PD	–	voltage test with PD measurement
DM	–	voltage test with dielectric measurement (mainly tanδ)
ACTC	–	transfomer circuit for AC voltage generation
ACRL	–	Inductance-tune resonant circuit for AC voltage generation
ACRF	–	frequency tuned resonant circuit for AC voltage generation
1)	–	applied in the past and no longer recommended
2)	–	mainly for PD diagnostics

6.2   Non-destructive electrical measurements

The principles involved in the non-destructive electrical test methods for monitoring and proving the condition of insulating materials in sample form and in equipment are outlined below. The majority of these methods are called for in various Standard specifications. Also included in Section 6.2.4 are more recent, well-developed techniques, which are being applied to determine their viability, especially for assessing the condition of insulation on site.

6.2.1   Insulation resistance (IR) measurements

The method measures the resistance of a dielectric or insulation structure when a high-value direct voltage is applied. The commonest instrument is the Megger, which consists of a hand- or motor-driven DC generator (D) – up to 5 000 V – and a built-in moving coil instrument for measuring the ratio ('resistance') of the applied voltage and the current flowing. The principle is depicted in Figure 6.10, in which are shown the current coil (1) and voltage coil (2). The torques produced act in opposite senses such that the meter deflection is dependent on the ratio of voltage to current. The scale is non-uniform, being more open at the higher-resistance values where the current is small.

Although low-resistance paths in series with good insulation are not easily detectable and parallel-path effects cannot be eliminated within structures, this simple test has been developed by the industry as a valuable assessment tool for indicating insulation condition in many applications [14]. It does not correctly monitor the AC operating conditions.

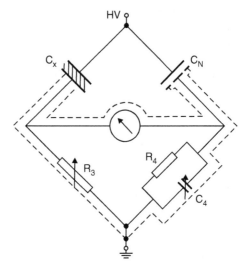

Figure 6.12   Principle of the high-voltage Schering Bridge

the testing transformer, careful screening of the low-voltage components and cables, as well as efficient earthing. For detailed discussions consult References 1 and 2.

6.2.2.2   Transformer ratio-arm bridge

For thirty years or more commercial transformer ratio-arm bridges have been available for measuring the DDF of insulation. These instruments are more sensitive, easier to screen against stray effects and more suitable for use on site than the HV Schering Bridge. Automated versions are offered for applications where many measurements are to be made, e.g. for oil testing. A circuit diagram of the principle of the bridge is given in Figure 6.13.

The main part of the bridge is a three-winding transformer or comparator. This has very low losses and leakage (high-permeability core) and is well shielded from stray magnetic fields. A major advantage is that no net mmf exists across windings W1 and W2 at balance. Also, it is unnecessary to allow for the stray capacitances of the low-voltage leads and those of the windings as no voltage appears across them – thus allowing the use of long leads [1]. As with the Schering Bridge a high-voltage standard capacitor C_N is required.

Allowance may be made for testing earthed and unearthed equipment. In the former case it must be possible to 'float' the lower end of the supply transformer. For the grounded condition two measurements are required – one with the test object disconnected (i.e. measurement of the stray capacitance to earth, C'_X) and the other with C'_X and arm C_X/R_X in parallel. The balance condition is achieved by varying the turns ratio of windings $1(N_1$ turns) and $2(N_2$ turns) to give zero output. At ampere-turn balance, $I_1 N_1 = I_2 N_2$, by adjusting R and C allowance can be made for the phase

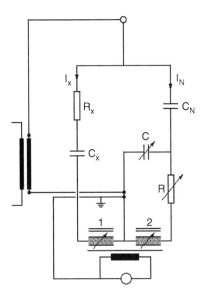

Figure 6.13   Transformer ratio-arm bridge for unearthed test object

shift in I_X due to the losses in the test object C_X. The latter is represented as a series circuit of R_X and C_X.

At balance with an ungrounded test object the values of C_x and $\tan\delta$ may be derived. The equations for determining the magnitudes of the relevant currents, neglecting the impedances of 1 and 2 and assuming an applied voltage of V, are as follows.

$$I_N = V\,j\omega C_N(1 + j\omega CR)/\{1 + j\omega(C + C_N)R\} \tag{i}$$

By division in parallel paths:

$$I_2 = I_N/(1 + j\omega CR) = V\,j\omega C_N/\{1 + j\omega(C + C_N)R\} \tag{ii}$$

$$I_1 = V\,j\omega C_x/(1 + j\omega C_x R_x) \tag{iii}$$

At balance $I_1 N_1 = I_2 N_2$ (iv)

From equations (ii), (iii) and (iv) and by equating real and imaginary parts:

$$C_x = C_N N_2/N_1 \tag{6.4}$$

$$\tan\delta = \omega R(C + C_N) \tag{6.5}$$

Because of the very low impedance of the windings it is possible to make measurements with a significant capacitance across winding 1. A value of 2000 pF is allowable in a particular test set. If this is unspecified, appropriate checks should be made, for example when measuring via the DDF tap of a bushing.

The standard capacitor, C_N, is usually a gas type (e.g. SF_6), having insignificant losses at full voltage and probably a capacitance of 100 pF. These units are very

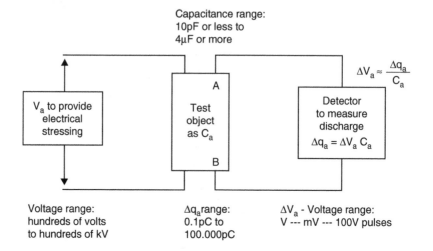

Figure 6.14   Block diagram showing PD measurement requirements

expensive. For conditions with an earthed test object, reference should be made to the manufacturer's instruction manual.

6.2.3   Measurement of partial discharges by electrical methods

The partial-discharge measurement requirements are indicated in the block diagram shown in Figure 6.14. For simplicity, it is assumed that the apparent charge pulse magnitude is the parameter to be measured The high voltage, V_a, is applied to the test object represented by an ideal capacitance, C_a. Partial discharges created by the stressing due to V_a will appear across the terminals AB as a small pulse voltage ΔV_a. The range of pulse voltages to be measured may be estimated approximately using the simple relationship (neglecting any parallel capacitance),

$$\Delta V_a = \Delta q_a / C_a \tag{6.6}$$

where Δq_a represents the discharge magnitude (0.1 pC up to 100 000 pC) expected in the tested equipment, the latter perhaps having capacitances in the range of 10 pF (cap and pin insulators) to tens of microfarads or larger in generators, power capacitors and cables. The pulses to be measured have rise times ranging from ns to several μs depending on the particular insulation structure.

The values of pulse voltages (ΔV_a) for a range of magnitudes for $C_a = 100$ pF and an inception voltage of $V_a = 10$ kV (RMS) are given in Table 6.2. The equivalent energy levels are also included. It should be noted that C_a represents the total effective capacitance across A and B.

It is clear that a large-ratio voltage divider is needed at the detector input if high test voltages are to be applied. Using a simple divider would produce an unacceptable loss in sensitivity due to attenuation of the small pulses. Fortunately, the applied voltage and the pulses to be measured have widely different frequency spectra – 0/50/300 Hz

Table 6.2   *Ideal PD energy levels and pulse voltages for applied inception voltage of 10 kV (RMS)*

For Δq_a	C_a	V_a (rms)	w (Joules)	ΔV_a
1 pC	100 pF	10 kV	0.7×10^{-8}	0.01 V
100 pC	100 pF	10 kV	0.7×10^{-6}	1 V
10,000 pC	100 pF	10 kV	0.7×10^{-4}	100 V

compared with hundreds of kHz up to several hundred MHz. By use of a high-voltage coupling capacitor this frequency difference enables a low-impedance path for the pulses to be retained between the terminals AB and the detector. At the same time it provides a high-ratio divider for reduction of the applied voltage by the addition of a high-pass filter at the detector input.

The form of the detector varies according to the parameter to be measured and many papers on the subject have been published. Probably the more important techniques can be grouped as follows:

(a) measurement of individual apparent discharge magnitude using a resistance or tuned circuit input;
(b) use of radio interference voltage (RIV) techniques with resistive input giving an 'integrated charge' value as a meter reading;
(c) measurement of a quadratic mean, the meter response being proportional to the integral of the squares of the discharge magnitudes;
(d) measurement of losses due to multiple-discharge sites (e.g. dielectric loss analyser).

Of these methods the first is probably the most widely used and accepted as a standard test; higher sensitivity is achieved using a tuned circuit input, because, among other advantages, the 50 Hz component can be reduced to a lower level. This technique and its application to practical cases will be described. Various commercial systems are now available but these differ only in their versatility and sensitivity of detection. The circuit developed by G. Mole of the ERA [28] will be used as the basis for the present discussion, but many of the concepts have been used by other investigators. Much of the review by Mason published in 1965 is still of relevance and includes a number of earlier references. The use of RIV measurements may be retained for some cases but it is clear that gross errors can arise and the readings may have little quantitative value as PD measurements. More advanced PD techniques developed for particular conditions are described in later chapters. These methods are not included in the revised version of IEC 60270 [S6/5]. The IEC has commenced (2006) the preparation of a new Standard covering unconventional partial-discharge measurements including UHF and acoustic techniques [S6/12].

(a)

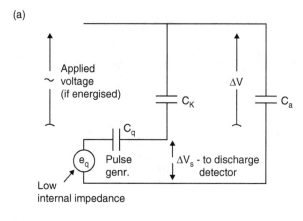

(b)

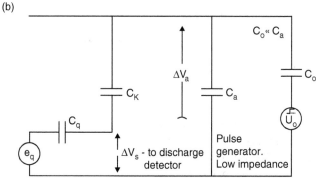

Figure 6.17   *Calibration of idealized capacitive circuits: (a) indirect (reference)*
calibrator; (b) direct calibrator

essential that the connection between the calibrator capacitor C_0 and the test object
be short and that the lead from the pulse generator to the capacitor be screened and
correctly matched. This requirement may introduce difficulties when testing large
(tall) components such as transformers, switchgear and bushings but is necessary in
order to obtain, as far as practicable, accurate and repeatable measurements under
adverse conditions. The external calibrator is usually battery-driven and should be
capable of calibrating the system for the expected high-voltage test PD values. Fixed
magnitudes of 5 pC, 10 pC, 100 pC and possibly 500 pC for transformer testing may
be appropriate. The requirement that $C_0/C_a \leq 0.1$ must be met.

The configuration in Figure 6.16 allows immediate response to changes in PD
levels during high-voltage tests by adjusting the amplifier gain and internal calibrator
settings, thereby retaining the original calibration. Linearity over the expected range
should be checked using an external calibrator. The alternative procedure (circuit) is
to determine the amplifier output in terms of pC/mm, pC/volt or a pC meter during
the calibration and maintain a fixed gain for the high-voltage test. It is possible to

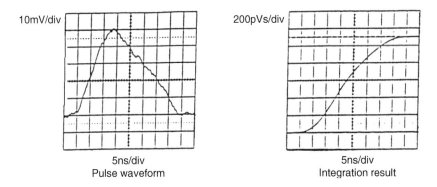

10mV/div

5ns/div
Pulse waveform

200pVs/div

5ns/div
Integration result

Figure 6.18   Example of calibrator check by integration method

use fixed attenuation ratios e.g. $\times 1$, $\times 10$, $\times 100$. In this system there is no direct comparison between the live discharge and the internal reference (e.g. e_q).

If a digital measuring system is in use the calibrator must be capable of producing pulses of adjustable repetition rate up to limits imposed by the resolution time of the system. Also the calibrator must be able to generate pulses at, say, 100 Hz for a specified time period in order to check that the digital instrument will record an equal number – a deviation of ± 2 per cent is allowed in IEC 60270. The latter contains an 'informative' Annex (E) giving guidelines for digital acquisition of the analogue partial discharge signals. A comprehensive review of the various problems and requirements is presented in Reference 19.

Calibration of the calibrator is essential in order to ensure acceptable accuracy during equipment tests. In Annex A of IEC 60270 it is specified that the charge shall be compared with a reference calibrator traceable to national standards. The comparison may be made with a PD measurement system or a digital oscilloscope with integration capabilities. A typical response is depicted in Figure 6.18 for a calibrator with $C_0 = 1$ pF, voltage 20 V and loading resistor 50 Ω. In this case there is good correspondence between the pC magnitude of the pulse injected, the area of the measured pulse (approximately) and the integrated value indicated by the digital oscilloscope.

For some years development has taken place in the application of inbuilt capacitor probes for measurement of PDs in GIS equipment. The technique has been validated in an *Electra* report [20] and is included as a calibration method in IEC 60270. See Chapter 9.

A simple method for determining the apparent measurement circuit sensitivity under live conditions is to connect across the test object a very low-capacitance PD source. Various techniques are applied including the temporary connection of a thin wire 'spike' at the HV terminal. As suggested by Kreuger [17] consistent discrete discharges at low-voltage can be obtained with a simple configuration of a needle mounted with its tip located at the centre of an earthed hollow hemispherical electrode of 25 mm internal radius. Discharge magnitudes of 20 pC or so at 1.7 kV can be achieved, depending on the point radius. Pulses of magnitudes of 25 to 250 pC for different point radii for a voltage of about 3 kV are quoted. This arrangement has been of value in making comparisons between different systems. PDs induced with thin wire or sharp points also allow confirmation of the position of the negative peak of the power frequency waves on the instrumentation display.

6.2.3.3   Reduction of interference

The reduction of interference occurring at the line terminal of a test object or plant item during a PD test can sometimes be minimized by use of a balanced circuit or a polarity discrimination circuit. Both configurations are included in IEC 60270 and are available commercially. The conventional ERA detector depicted in Figure 6.16 is designed to operate from a centre-tapped transformer with dual inputs suitable for connection in a balanced circuit and is capable of effecting rejection ratios of 30 to 100. Another well-established equipment developed by Kreuger and used for many years has two separate adjustable impedances allowing rejection ratios of up to 1 000 under favourable conditions [17]. The discrimination circuit developed by Black [21] recognises the polarity of the pulses at the detecting impedance. If those occurring simultaneously are of identical polarity they are rejected. These instruments have been applied in switch-yards up to 300 kV. An earlier CIGRE publication [22] may be of assistance in understanding some of the interference problems associated with PD detection. More advanced techniques and recent developments for eliminating or reducing interference are discussed in Chapter 9.

6.2.3.4   Presentation of PD measurements

In the earlier detectors – for example as developed by Mole and later by Kreuger – a visual oscillographic display was incorporated to allow observation and measurement of the amplifier output and for determining the approximate time of occurrence within the power-frequency cycle. Additionally a meter reading of the largest pulse was included.

This relatively simple concept and effective system – with some minor changes – has been the industry norm for 50 years. However, in the 1960s minicomputers made it possible to record and analyse data [23] that, previously, were recordable only by bulky PHA systems or directly from the oscilloscope screen.

The new systems became more practical following development of microcomputers until, in the 1980s, the new PCs resulted in an upsurge of portable computer-based systems in the 1990s. The latter instrumentation allows complex analyses to be carried out although the greatest value is probably the ease with which trending of changes in maximum and average charge, repetition rates and pulse-power variations can be recorded during a test and for future reference. Many attempts at relating the measurable PD characteristics to the conditions within particular insulation systems – as in the past [24] – have been made [25,26] but it is not clear that this can be achieved for all circumstances [27]. This reference also supports the view that for oil-impregnated materials the magnitude, phase relationships and repetition rates of the PDs may be more significant than the more complex parameters that can be calculated from the basic data. An example of a computer record for a discharging source is given in

Figure 2.13. CIGRE Brochure 226 [26] presents the probable response for a wide range of conditions involving partial discharges in samples and equipment. Interestingly, a number of faults in substations have been identified by excessive interference on portable telephones.

6.2.4   *Dielectric response measurements*

The application of polarization/depolarization measurements for assessing the quality of insulating materials is a relatively simple procedure but analysis of the data is very complex. In the following an indication is given of the time and frequency domain methods currently available. Details of the various analyses relating measurements to moisture content, ageing and possible contamination will be found in the references. A number of significant parameters can be measured at well below operating values. One of the earliest instruments was that developed by Mole [28] at the ERA in the early 1950s. The *dispersion meter* was used for many years for indicating the moisture levels in oil–paper systems, especially transformers. The advances in technology and interpretation of dielectric theory have allowed the development of improved methods.

Possibilities for time- and frequency-domain measurement have been investigated in detail during the past ten years and a number of instruments and experimental systems developed. Many tests are being carried out on equipment with oil-impregnated insulation structures, some on XLPE cables and a few on surge arresters associated with HV proving tests [29]. A review of the theory and background of dielectric response measurements is given by Zaengl [30], including 103 references. The particular case for transformers is considered in a report by CIGRE TF 15.01.09 [31].

In the following are outlined the parameters involved and the principles of measurement. The various instruments and special test procedures are not described, as the newer techniques are being trialled and assessed with respect to their value for monitoring particular apparatus.

6.2.4.1   Time-domain measurements

Low-value direct voltages are applied to the test object for a specified time t_c followed by a chosen short circuit period before isolating the sample. Depending on the test procedure it is possible to measure the voltage or currents as indicated in Figure 6.19. Many on-site tests have been carried out attempting to relate these variations to the insulation condition of oil–paper transformers, current transformers and cables (oil–paper and XLPE) using experimental methods and commercially available instruments. The techniques are not universally accepted and caution in interpretation of results is necessary, especially for transformers containing oil gaps in series with the cellulosic insulations [31].

An advantage of the procedures is that for known conditions significant data can be obtained for low electric stresses on completed equipment. The tests are applied offline and are very lengthy, and care must be taken to minimise interference, as the quantities being measured are small.

of the number of voids involved and the ageing stage of the insulation, but care in interpretation is essential [30]. A form of this method is being applied commercially.

6.2.4.2   Dielectric frequency domain spectroscopy measurements (FDS)

A number of investigations have been completed in order to determine whether capacitance and loss changes over a range of frequencies (e.g. 0.0001–1 000 Hz) might indicate the presence of moisture in transformers and also in XLPE cables It is possible this technique may be applicable to AC cables at moderately high stresses. An application of the technique is given in Reference 34, where a frequency range of 0.1 to 10 Hz was used up to 30 kVpk.

6.3   Physical and chemical diagnostic methods

There are many non-electrical techniques for determination of the condition of the insulating materials used in power equipment. The predominant application is for oil and cellulosic materials, although new techniques are being developed for assessing the state of, for example, composite insulators, the by-products of SF_6 decomposition and aged XLPE cables. A major restriction to application of the methods is the lack of access to the materials when built into equipment. Samples of oil in the case of oil-filled transformers, switchgear, cables, bushings and of gas from GI equipment may be obtained without harmful effects. However, the removal of solid insulation is normally impossible because of resulting damage to the insulation structure of the apparatus.

6.3.1   Indicators of in-service condition of oil–paper systems

In-service indicators of the condition of the oil-impregnated materials within transformers, tap changers, bushings and cables include the levels of concentration of moisture, gas, furans and particles in the oil as well as its overall quality. During refurbishment of power transformers it may be possible to remove paper/pressboard samples from non-critical locations for determination of their moisture content and degree of polymerization.

6.3.1.1   Moisture content in oil-impregnated systems

A significant factor in the operation of oil-impregnated paper systems is the need to maintain low moisture contents. The simplest technique is to monitor oil samples by means of the well-established Karl Fischer method (IEC 60814, S6/6). This publication includes tests for oil-impregnated paper and pressboard. Care must be taken when removing the required oil samples. The oil temperature at the time of removal should be noted, together with the type of oil-cooling system if operative. Preferred techniques are described in IEC publications 60475 [S6/7] and IEC 60567 [S6/8].

The oil-impregnated cellulosic material moisture levels are estimated from the oil values as indicated in Figure 6.20. These equilibrium curves were developed by Oommen in 1983. A good review of the variables and history of the derivation of

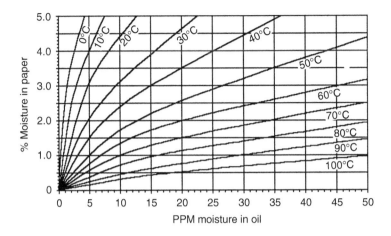

*Figure 6.20   Oil–paper moisture equilibrium curves (after Oommen) [35] [repro-
duced by permission of the IEEE]*

relationships used by the industry, including the original research by Fabre and Pichon (1960), are presented in reference [35]. Moisture contents are much less than 1 per cent at operating temperatures for new transformers through to 2–3 per cent for a well-maintained unit (with silica breather) and perhaps 4–5 per cent after thirty years of service. The last value can lead to rapid ageing and possible failure.

If the moisture in the paper and oil are not in equilibrium and not at the same temperature, errors in the estimate will arise but in practice the technique is found to be valuable.

Direct measurement (IEC 60814, S6/6) of paper/pressboard samples from failed units – e.g. current transformers – are often carried out during 'postmortem' investigations and refurbishments. As with oil samples, care must be taken during removal and records made of the length of time of exposure to air (and its humidity), the unit history and the precise location within the structure.

6.3.1.2   Dissolved gas in oil measurements

DGA is probably the prime online monitoring technique for detection of developing faults in power transformers and has been applied for more than thirty years. The gases considered to be of significance are hydrogen (H_2), methane (CH_4), acetylene (C_2H_2), ethylene (C_2H_4), ethane (C_2H_6), carbon monoxide (CO) and carbon dioxide (CO_2). Values vary from a few ppm to hundreds for the hydrocarbons and from several hundreds to thousands for the carbon monoxide/carbon dioxides. These levels are measured in the laboratory – or possibly within a mobile unit – using gas chromatographs manufactured for this specific function. The procedures are recommended in IEC 60567 [S6/8]. In order to obtain reliable and repeatable results, it is essential that the oil sampling be carried out according to a prescribed protocol [S6/8] but related to the particular equipment design: power transformers, tap changers, bushings, instrument transformers or cables.

A number of online monitoring devices are available, in particular for detecting changes in the hydrogen levels in power transformers (Chapter 9).

Over the years various methods of interpretation have been proposed for identifying faults in the insulation structure – in particular the ratio system, one form of which is presented in IEC Publication 60599 [S6/9]. The coded list of faults related to particular practical situations is discussed by Duval and dePablo in Reference 36 and extended in 37. The databases and analyses are based on extensive studies by IEC TC10 and CIGRE Task Forces of SC 15. This work was continued to cover cases for which existing criteria may give an unsatisfactory interpretation. References 38 and 39 include extracts covering 179 cases where faults were identified by inspection and related to the gas levels. Also tabulated are 'normal' values for gases to be expected in satisfactory power and instrument transformers. It appears that the magnitudes are considerably higher than those allowed by many utilities before additional monitoring is required. Although there are some difficulties in interpretation, the technique gives a good indication of possible problems, especially if applied periodically as a trend measurement based on the experience of the particular authority. Examples from one source are presented in Chapter 9 and an advanced form of dissolved-gas analysis utilizing fuzzy-logic methods described in Chapter 10.

6.3.1.3   Furan measurements

The technique utilizing high-performance liquid chromatography for detecting furans in insulating oils is well established and is being applied by many utilities and manufacturers worldwide. The level of concentrations of different forms of furans has been shown to be indicative of the deterioration of cellulose at normal operating temperatures and may be related to anticipated life under certain conditions. The concentrations of significance are of the order of 0.05 mg/L($\approx$ppm) through to 5 mg/L($\approx$ppm) or higher for badly degraded insulation [38,39].

6.3.1.4   Particle counts

The concentration of different types and dimensions of particles in oil can be quantified by use of such procedures as those specified in Standard S6/10 (see under Section 6.6 below).

6.3.1.5   Oil quality

Many tests are specified for ensuring insulating oils are of the quality required. A number of Standards within the IEC describe the requirements and test methods, in particular IEC 60296 [S6/11] and associated publications. The methods covered include:

- breakdown tests
- oxidation determination
- permittivity, DDF and resistivity measurements

- sampling methods
- water determination

ISO Standards include determination of

- viscosity
- flash point
- pour point

In transformer practice the more important characteristics as indicators of oil deterioration are often considered to be acidity, DDF and interfacial tension [14].

6.3.1.6 Degree of polymerization

Oil-impregnated paper samples removed during refurbishing can be checked for their degree of polymerization. The temperature profile within a transformer may be determined from such measurements as well as determining the probable life expectancy of the insulation. DP values are in the range 1 000 for new units to 200 for those near their end of life [38,39]. Problems have arisen in obtaining comparable results between laboratories.

6.3.1.7 Other techniques

The condition of samples of deteriorated materials may be examined by many techniques including IR, X-ray, SEM, UV (for detecting waxed state) and molecular changes [40].

6.3.2 *Analysis of SF$_6$ samples from GIS*

Samples removed from a GIS are analysed by gas chromatography and mass spectrometry. The objective is to detect any SF$_6$ decomposition products that might have been created by, for example, low-level partial discharges.

6.3.3 *Surface deterioration of composite insulators*

Although the usual method of determining possible surface deterioration of composite insulators is by visual observation and perhaps leakage-current measurements, a technique has been developed [41] in Australia in which slivers of the surface material are removed for detailed quantitative analysis. By use of a special 'hot stick' the method has been successfully applied under live-line working conditions.

6.3.4 *Water treeing in XLPE cable insulation*

Application of the newer non-destructive techniques such as described in Section 6.2.4. has increased the possibility of detecting water trees in plastic cables. Following failures or suspected faults the only effective method of investigation is to cut out lengths of the cable and from suspected regions carefully prepared thin wafers for physical inspection. These methods are still required to confirm the presence of trees and the state of the dielectric, although new techniques are being developed.

19. CIGRE SC 33 TF 33.03.05, 'Calibration procedures for analog and digital measuring instruments', *Electra*, October 1998;(180):123–44

20. CIGRE SC 15.03 TF 15.05.02, 'Partial discharge detection system for GIS: Sensitivity verification for the UHF method and the acoustic method', *Electra*, April 1999;(183):75–87

21. Black, I.A., 'A pulse discrimination system for discharge detection in electrically noisy environments', *Proceedings of the International HV Symposium*, Zürich, September 1975

22. CIGRE, 'Elimination of interference in discharge detection', *Electra*, 1977;(21)

23. Austin, J., and James, R.E., 'On-line digital computer system for measurement of partial discharges in insulation structures', *IEEE Transactions on Electrical Insulation*, December 1976;**11**(4):129–39

24. CIGRE Working Group 21.03, 'Recognition of discharges', *Electra*, December 1969;(11):61–98

25. James, R.E., and Phung, B.T., 'Development of computer based measurements and their application to PD pattern analysis', *IEEE Transactions on Dielectrics and Electrical Insulation*, October 1995;**2**(5):838–56,

26. CIGRE WG D1.11, 'Knowledge rules for partial discharges in service', *Electra*, April 2003;(207):63–6 (CIGRE Brochure 226)

27. Lundgaard, L.E., *et al.*, 'Partial discharges in transformer insulation', CIGRE Task Force 15.01.04, Paper 15-302, Paris, 2000

28. Mole, G., 'Improved test methods for the insulation of electrical equipment', *Proceedings of the Institution of Electrical Engineers*, Section IIA, no. 3, 1953

29. Mardira, K.P., Saha, T.K., and Sutton, R.A., 'Investigation of diagnostic techniques for metal oxide surge arresters', *IEEE Transactions on Dielectrics and Electrical Insulation*, February 2005;**12**:50–59

30. Zaengl, W.S., 'Dielectric spectroscopy in time and frequency domain for high voltage power equipment' *IEEE Electrical Insulation Magazine*, 2003: Part 1, September/October;**19**(5):5–19; Part 2, November/December;**19**(6):9–22

31. CIGRE Task Force 15.01.09, 'Dielectric response methods for diagnostics of power transformers', *IEEE Electrical Insulation Magazine*, May/June 2003;**19**(3):12–18 (see also *Electra*, June 2002;(202):25–36)

32. Bognar, A., Kalocsai, L., Csepes, G., Nemeth, E. and Schmidt, J., 'Diagnostic tests of h-v oil–paper insulating systems (in particular transformer insulation) using dc dielectrometrics', CIGRE Paper 15/33-08, Paris 1990

33. Saha, T.K., and Purkait, P., 'Investigation of polarization and depolarization current measurements for the assessment of oil–paper insulation of aged transformers', *IEEE Transactions on Dielectrics and Electrical Insulation*, February 2004;**11**:144–54

34. Hvidsten, S., Holmgren, B., Adeen, L., and Wettstrom, J., 'Condition assessment of 12- and 24-kV XLPE cables installed during the 80s. Results from a joint Norwegian/Swedish research project', *IEEE Electrical Insulation Magazine*, November/December 2005;**21**(6):17–23

35. Du, Y., Zahn, M., Lesieutre, B.C., Mamishev, A.V., and Lindgren, S.R., 'Moisture equilibrium in transformer paper/oil systems', *IEEE Electrical Insulation Magazine*, January/February 1999;**15**(1):11–20
36. Duval, M., and dePablo, A., 'Interpretation of gas-in-oil analysis using new IEC Publication 60599 and IEC TC 10 databases', *IEEE Electrical Insulation Magazine*, March/April 2001;**17**(2):31–41
37. Duval, M., 'A review of faults detectable by Gas-in-Oil analysis in transformers', *IEEE Electrical Insulation Magazine*, May/June;**18**(3):9–17
38. dePablo A., 'Interpretation of degradation models of furanic compounds', CIGRE WG 15.01 TF 03 (1997) (see also *Electra*, 1997;(175))
39. Oommen, T.V., and Prevost, T.A., 'Cellulose insulation in oil-filled power transformers: Part II – Maintaining insulation integrity and life', *IEEE Electrical Insulation Magazine*, March/April 2006;**22**(3):5–14
40. Hill, D.J.T., Le, T.T., Darveniza, M., and Saha, T.K., 'A study of degradation of cellulosic insulation materials in a power transformer – Part 1: Molecular weight study of cellulose insulation paper', *Polymer Degradation and Stability*, 1995;**48**(1):79–87
41. Birtwhistle, D., Blackmore, P., Krivda, A., Cash, G., and George, G., 'Monitoring the condition of insulator shed materials in overhead distribution networks', *IEEE Transactions on Dielectrics and Electrical Insulation*, October 1999;(6):612–19
42. Harrold, R.T., 'Acoustic techniques for detecting and locating electrical discharges', *Engineering Dielectrics*, vol. 1, Chap. 10 (ASTM Technical Publication 669, 1979) (see also *IEEE Trans. Electr. Insul.*, vol. 11, 1976)

6.6   Standards related to basic test methods

S6/1 IEC 60052 (Ed. 3.0, 2002): Voltage measurement by means of standard air gaps
S6/2 IEC 60060: High-voltage test techniques:

Part 1 (Ed. 2.0, 1989): General definitions and test requirements
Part 2 (Ed. 2.0, 1994): Measuring systems
Part 3 (Ed. 1.0, 2006): Definitions and requirements for on-site testing

S6/3 IEC 62067 (Ed. 1.1 2006): Power cables with extruded insulation and their accessories for rated voltages above 150 kV ($U_m = 170$ kV) up to 500 kV ($U_m = 550$ kV) – Test methods and requirements
S6/4 IEC 61083-1 (Ed. 2.0, 2001): Instruments and software used for measurement in high-voltage impulse tests – Part 1: Requirements for instruments
S6/5 IEC 60270 (Ed. 3.0, 2000): High-voltage test techniques – Partial discharge measurements
S6/6 IEC 60814 (Ed. 2.0, 1997): Insulating liquids – Oil-impregnated paper and pressboard – Determination of water by automatic coulometric Karl Fischer titration
S6/7 IEC 60475 (Ed. 1.0, 1974): Method of sampling liquid dielectrics

S6/8 IEC 60567 (Ed. 3.0, 2005): Oil-filled electrical equipment – Sampling of gases and of oil for analysis of free and dissolved gases – Guidance

S6/9 IEC 60599 (Ed. 2.0, 1999): Mineral-oil-impregnated electrical equipment in service – guide to the interpretation of dissolved and free gases analysis

S6/10 ASTM D6786-0 2: Standard test method for particle count in mineral insulating oil using automatic optical particle counters

S6/11 IEC 60296 (Ed. 3, 2003): Fluids for electrotechnical applications – Unused transformers and switchgear

S6/12 IEC 62478 (Ed. 1.0): High-voltage test techniques: Measurement of partial discharge by electromagnetic and acoustic methods (new project, 2006)

6.7   Problems

1. Discuss the accuracy problems when applying (i) resistive and (ii) capacitive dividers for measurement of high-voltage lightning impulses.

   A resistive unit has a high-voltage arm of $100\,k\Omega$ and a low-voltage arm of $50\,\Omega$. It is connected to measuring instrumentation by a $75\,\Omega$ coaxial cable. How should the cable be terminated and what is the HV/LV measurement ratio for a $1.2/50\,\mu s$ impulse wave?

   For a capacitive divider of $100\,pF$ HV and $0.2\,\mu F$ LV arms, determine the ratio and appropriate cable terminations.

2. (i) Indicate why care must be taken when interpreting the readings from a Megger-type IR measurement.

   (ii) Outline the advantages of a transformer ratio arm bridge compared with a Schering Bridge for measurement of insulation loss angles on power equipment.

3. Derive the relationships for the loss angle and capacitance values of the unknown capacitor C_x in the Schering Bridge circuit in Figure 6.12.

   During a high-voltage test on new oil-impregnated bushing the permanent values of the Bridge components were $C_N = 100\,pF$, $R_4 = 318\,\Omega$. At balance $R_3 = 700\,\Omega$ and $C_4 = 0.086\,\mu F$. The test frequency was $50\,Hz$.

   Determine the bushing capacitance and loss angle. Comment on the $\tan\delta$ value if the test was carried out at $1.05\times$ operating voltage to ground. Refer to S7/18 of Chapter 7. What further action might be taken?

4. High voltage partial discharge tests are to be applied to a range of power capacitors. Detail the test and measuring equipment required including a $100\,pF$ coupling capacitor.

   Determine the relative sensitivities for tests on the following capacitors. (i) $50\,pF$, (ii) $1000\,pF$ and (iii) $10,000\,pF$. If the desired sensitivity for (iii) cannot be achieved how might the test circuit be modified?

Chapter 7

Established methods for insulation testing of specific equipment

- Type tests, routine tests and special tests
- Insulators, surge arresters, switchgear, bushings, capacitors, cables, HV rotating machines and transformers
- Insulation tests for HVDC components

This chapter is concerned with existing acceptance and testing procedures for assessing the condition of the insulating materials and insulation structures in completed high-voltage power-system equipment. The techniques (see Chapter 6) chosen for proving the satisfactory state of a particular item will depend on the form and magnitude of the test voltage required, the physical size of the component, its complexity, the loading on the test supply circuits and the monitoring/fault-detection methods adopted.

Tests may be designated as *type* (design), *sample* or *routine*, depending on the form and number of items manufactured. In some instances a *prototype* is produced, which is subjected to a wide range of tests before full-scale production begins. Testing scenarios vary widely with the type of product. For example, cap-and-pin insulator samples of four or five units may be chosen for test from a batch of 2 000 while, with large plant such as generators and transformers, a series of type, routine and commissioning tests are carried out on the 'one-off' completed unit.

The test methods and the characteristics to be checked for each type of equipment are usually based on international and national standards formulated by experts in the particular field and subsequently approved by representatives of users, manufacturers, testing laboratories and research organizations. Many of these publications contain valuable technical information and application guides. Information on the wide range of Standards available may be obtained by contacting IEC and various national organizations such as the IEEE, ASTM, BSI, VDE and SA (see Appendix 2). In some fields new nonstandardized techniques are being trialled by the industry.

at least half the length of the external flashover path (puncture-proof). For units with the solid length less than half of the total, and for the pedestal-post type in which the thickness of the solid is small compared with the external distance, electrical sample and routine tests are specified. References S7/4 and S7/5 are relevant.

Type tests are applied to one insulator only. The procedures are as in 7.1.1.

For indoor applications, dry lightning impulse, switching impulse and power-frequency withstand-voltage tests may be required. For outdoor applications, wet switching impulse and power-frequency tests may be required. The relative test levels for particular lightning-impulse values are detailed in Reference S7/4. Because of the variability in the basic insulation levels the dry lightning-impulse value is chosen as the reference and not the system-highest voltage. Note that switching impulses (250/2500 μs) would usually be specified for units intended for operation in systems with highest voltages of 300 kV and above.

Sample puncture tests may be required on units where the solid insulating material distance is half or less than that of the flashover path.

Routine electrical tests are specified for the same designs as the sample tests.

7.1.3   Composite insulators for overhead lines (string and post units)

Extensive *design tests* are called for in the assessment of composite insulators. These involve dry power-frequency flashover tests for both the line [S7/6] and post [S7/7] insulators, in addition to steep fronted impulse (1 000 kV/μs) applications for the special conditions specified. Tracking tests are also required. The comprehensive sequence of mechanical, thermal and moisture tests (including leakage current measurements) may be found in references S7/6 and S7/7. *Type tests* include dry lightning impulses, wet switching impulses and wet power-frequency withstand voltages. These are applied to one insulator or insulator unit only. No electrical tests are specified for sample checks and for routine inspections.

Within the Standard, voltage levels are given for an optional radio-interference voltage test (RIV). No RIV values are quoted but the test voltage indicates the value at which corona would be expected to extinguish under dry conditions. Acceptable levels are agreed between the purchaser and manufacturer – a guide for establishing limit values is given in CISPR 18-2, Amendment 1 [1]. Detailed procedures for obtaining RI characteristics for dry high-voltage insulators are given in IEC 60437 [S7/8]. The specified reference frequency given in CISPR 18-2 is (0.5 ± 0.05) MHz but others in the range 0.5–2 MHz may be selected based on previous practice. It is expected that RI disturbances from insulators will not normally affect television reception.

7.2   Overhead line and substation hardware

Although the Standards recommend air clearances for the design and operation of overhead line systems and substations, it is often necessary to prove the performance of a particular configuration, especially at the higher voltage levels, where switching-impulses or unusual pollution conditions exist. Such testing includes the determination of corona inception levels in order to ensure that unexpected flashovers do not occur and that interference does not affect local radio transmissions and receptions. This interference can invalidate some of the monitoring systems being used on site for the detection and measurement of partial discharges.

In the laboratory, levels can be quantified by PD measurements and the Radio Interference Voltage methods described in CISPR 18-2. Usually, insulators and the HV hardware will be checked simultaneously. The measurements are made utilizing a standard CISPR test set as specified in CISPR 16-1:1993 [2]. The technique has been applied for many years and is sometimes used as a partial-discharge test (see IEC 60270).

The diameters of overhead conductors for a given voltage and condition are well defined, as in Reference 3. Sometimes, however, it is necessary to determine the self-screening efficacy of multiple conductors and the satisfactory performance of a new design of spacer (or spreaders), as in quad arrangements. Proposed electrode configurations at the terminals of bushings, switchgear and arresters as well as insulators will require checking in the high-voltage laboratory at above operating voltages requiring appropriate corona-free supply sources.

7.3 Surge arresters

Surge arresters are a vital part of the power system. They must be shown to be able to withstand the steady-state voltages without deterioration, to respond correctly by reducing an excessive lightning or switching surge to a safe value and then to recover their previous power-frequency strength.

The earlier units included spark gaps and silicon carbide elements requiring tests as detailed in IEC 60099-1 [S7/9]. The tests for the later designs of metal-oxide gapless units are described in Part 4 of the Standard. The testing procedures are complex and extensive. In the following are summarized some of the type and routine tests for proving of the insulation requirements and V-I performance.

Type tests on the arrester housings include a lightning-impulse dry withstand-voltage test and a similar switching-impulse test both above the corresponding protection level. In the latter case a wet test is required for outdoor units. The power-frequency withstand test is applied for one minute with a peak value of just below the lightning-impulse protection level.

Residual-voltage-type tests on three complete sample arresters determine the highest residual voltage at the rated discharge current. From the results of other tests the switching impulse protection level for specified currents may be defined.

Type tests are carried out in order to simulate current flow from a recharged transmission line. Long-duration current-impulse withstands with virtual lengths of the wave peak within the range 500–3 000 μs are applied. A simplified circuit is described in Appendix J of IEC 600994.

Operating-duty-type tests incorporating the simultaneous application of impulse and power-frequency voltages are also included in the Standard. A major aim is to prove that thermal runaway does not occur and that the unit can cool down during the power-frequency periods. A special multi-pulse operating-duty test representing multiple lightning strokes has been proposed [4, S7/10]. This has been shown to fail some units that would have passed the normal requirements.

Routine reference voltage tests require determination of the power-frequency voltage necessary to produce the reference current as measured at the bottom of the unit. The voltage value must be within the range specified by the manufacturer. This is a parameter that may be used for monitoring in service.

Routine lightning-impulse residual-voltage tests are carried out at, preferably, the nominal discharge current. The voltage must not exceed the appropriate specified values for a complete arrester – typically of the order of 3 × the rated line voltage.

As part of the routine tests a partial-discharge contact noise measurement is made. The internal partial discharges, including disturbances due to bad contacts, should not exceed the equivalent of 10 pC. It may be practical to monitor the condition of suspect units in service by means of PD measurements.

7.4 Switchgear

In this section switchgear is deemed to include circuit breakers for on-load switching and fault protection, disconnectors (operating off-load) for isolation of the high-voltage circuits, metal-enclosed switchgear and on-load tap changers installed in transformers for changing the system voltage levels. For all of these devices a special requirement is the proving of the dielectric strength across open contacts, in addition to the usual insulation overvoltage checks to ground and between phases.

7.4.1 Circuit breakers

Extensive electrical, thermal and mechanical tests are specified for type and routine testing of circuit breakers [IEC 60694, Ed. 2.2 B, and 62271-100, Ed.1.0 B, Refs S7/11 and S7/12]. The primary dielectric tests are summarized below. The power-frequency, lightning-impulse and switching-impulse withstand test voltages are tabulated in the specifications.

For ≤ 245 kV class the phase-to-phase, phase-to-earth and open-switch withstand test voltages (one-minute power frequency and 1.2/50 μs impulse) are equal, with the values for the isolating distances approximately 10–15 per cent higher.

Above 245 kV class the power-frequency test value across the open contacts is higher than the phase-to-earth levels; the switching-impulse test level between phases is greater than the value to earth; and, also, a combined switching-impulse/power-frequency test is required for checking the isolating distances. A combined lightning-impulse/power-frequency withstand-voltage test is applied across open contacts/isolating distances.

One method for applying 'longitudinal' test voltages without exceeding the withstand level to earth is by application of two separate out-of-phase power-frequency sources and, for impulse tests, an impulse of the withstand value to earth to one terminal with a complementary voltage of opposite polarity to the other terminal. If a second impulse-voltage source is not available, the peak of the power frequency may be utilized. The technique is described in IEC 60060-1.

If a second voltage source is not available for open-switch or isolating distance tests, one voltage may be applied, but the value from the tested terminal to the frame of the device must not exceed the normal withstand level. The required longitudinal test voltage is achieved by increasing the frame potential to above earth and earthing the other terminal of the open contacts.

The high-voltage laboratory or factory test facility must be capable of carrying out these various procedures, the majority of which are type tests only.

7.4.1.1 Type tests

For 245 kV-rated voltage and below:

- power-frequency withstand-voltage tests for one-minute dry (indoor), wet (outdoor);
- lightning-impulse withstand-voltage tests with voltages of both polarities (1.2/50 μs) – dry tests only.

For above 245 kV-rated voltage:

- power-frequency voltage withstand tests for one-minute dry tests only;
- switching-impulse withstand-voltage tests with voltages of both polarities (250/2 500 μs) – dry (indoor), wet (outdoor);
- lightning impulse withstand-voltage tests with voltages of both polarities (1.2/50 μs) – dry tests only;
- Partial-discharge tests;
  (PD tests may be specified as agreed between manufacturer and customer. The tests are of importance for GIS equipment for which the standard method (see Chapter 6) or the UHF measurement systems may be used (see Chapter 9).)
- radio interference voltage test (RIV).
  (RIV tests may be specified for equipment rated at 123 kV and above. The measurements are made with the contacts in both open and closed states with all other terminals earthed. The procedures for switchgear are detailed in IEC 60694. An RIV versus applied voltage characteristic curve is recorded. The acceptance level is 2 500 μV at 1.1 $\times$ phase-to-earth voltage. No guidance is given as to possible equivalent levels measured by conventional PD test methods. Such information could be of value when monitoring equipment in substations.)

7.4.1.2 Routine tests

Short-term power-frequency withstand tests are specified for new dry conditions on the complete apparatus or individual terminals. If solid-core insulators provide the major insulation, the high-voltage tests may not be necessary.

Routine dry, power-frequency, one-minute, withstand-voltage tests are applied to all bushings and partial-discharge tests at just above rated voltage. If required for a transformer installation the voltage is increased to $1.5 \times$ rated voltage. PD acceptance values are 10 pC for these conditions except for resin-bonded and cast-resin insulation systems, where levels of 100/300 pC and 100 pC respectively are usually taken as reasonable limits.

The insulation of the bushing test tap must be checked by an applied voltage of 2 kV and the DDF and capacitance values measured. These values should be 100 mR and 5000 pF or less. The lower the tap capacitance the more sensitive is the PD test as carried out according to IEC 60270 for transformers (7.10.2) and for some switchgear.

When testing for special applications it is important to use correct terminations (e.g. for re-entrant types) and to ensure that clearances within the test tank are typical of those in the equipment in which the bushing is to be installed. Air terminations may need to be screened, rod gaps removed and attention be given to the method of mounting/clamping the earthed flange to the test tank. Temporary clamps can be a source of corona. The measurement of low-value PDs requires the usual 'good housekeeping' and preferably, for routine tests, a dedicated area where the PD circuit layout can be permanent.

Guidance is given in IEC 61464 [S7/19] regarding the interpretation of DGA results for in-service oil-impregnated-paper bushings. Normal site tests include DDF and IR measurements.

7.6 High-voltage instrument transformers

The tests required for the wide range of instrument transformers applied in the power systems are detailed in IEC 60044 Parts 1–8, which includes current transformers (CTs), voltage transformers (MVTs and CVTs), combined transformers (CV/VT), and electronic types. The developments in instrument transformer design have necessitated updates of the standard.

7.6.1 Current transformers

The high-voltage withstand tests on CTs are applied to the paralleled primary winding terminals with the outer screens and secondary windings earthed. The major dielectric tests are summarized below. The test voltage values are based on those in IEC 60071 and tabulated in IEC 60044-1 [S7/20].

7.6.1.1 Type tests

Lightning-impulse voltage-withstand test: For $U_m < 300$ kV. A maximum of 15 standard impulses of each polarity is to be applied and for $U_m \geq 300$ kV only three standard impulses of each polarity are required. Waveshape changes may indicate failure.

Switching impulse-voltage withstand test: For $U_m \geq 300$ kV 15 switching impulses of positive polarity are to be applied. The test is to be under wet conditions for outdoor units. The failure criteria are similar to those for the U_m < 300 kV lightning impulse tests.

Wet power-frequency withstand-voltage test: For U_m < 300 kV the test voltage is to be applied for 60 seconds.

Radio-interference voltage measurement: RIV tests are carried out on CTs for use in air-insulated systems with U_m ≥ 123 kV. During the test hardware is required to simulate operating conditions. Refer to IEC 60044, Part 1, for details. By agreement a PD test may be substituted during which the measured value must not exceed 300 pC. It should be noted that the RIV and PD magnitudes are not directly related.

7.6.1.2 Routine tests

Dry power-frequency withstand voltage: The test voltage is applied for 60 seconds between the shorted primary terminals and earthed shorted secondary windings/frame. The appropriate test levels are tabulated in IEC 60044-1.

Partial-discharge measurement: Partial-discharge tests are applicable to units of ≥ 7.2 kV. The test method is specified in IEC 600270. A PD level of 5 pC should be detectable. The test may be performed with one of the two procedures: (i) during reduction of the withstand voltage; (ii) after the withstand test by pre-stressing for 60 seconds at 0.8 × withstand voltage followed by reduction to the PD test voltage.

The PD test voltage is maintained for 30 seconds. The allowable levels at U_m and 1.2 $U_m/\sqrt{3}$ are 10 pC and 5 pC respectively for liquid-immersed units and 50 pC and 20 pC for solid types. The system neutral is to be earthed.

7.6.1.3 Special tests

Chopped impulse tests on primary winding: Negative standard lightning impulse test voltages with chopping times between 2 μs and 5 μs are applied for U_m < 300 kV and U_m ≥ 300 kV conditions. Variations in the full-wave (FW) shapes before and after the chop-wave (CW) applications may represent an internal fault.

Multiple chopped impulse test (U_m≥ 300 kV): Note: This test is included as an informative Annex (B) of IEC 60044-1 proposing a possible procedure for determining the response to high- frequency stresses as may be produced by, for example, isolator switching. Dissolved-gas-in-oil analyses (IEC 60599) before and after the test, with a three-day interval, are suggested for monitoring failure.

Capacitance and dielectric dissipation factor (DDF) measurements: These measurements apply only to liquid-immersed units for U_m ≥ 72.5 kV. Values in the range 10 kV to $U_m/\sqrt{3}$ are to be recorded. The aim is to check the quality of production. Measurements are typically <0.005, possibly 3 mR or less.

7.6.2 Inductive voltage transformers

The dielectric test requirements for inductive-type transformers are considered separately from the capacitive configuration. The relevant standard is IEC 60044-2 [S7/20]. All the specifications in relation to insulation tests are similar to those in the Current Transformer specification with the following exceptions.

7.6.2.1   Impulse withstand tests

During the impulse tests records are required of earth currents and of voltages transferred to the secondary windings. These are necessary in order to improve the sensitivity of fault detection as in the procedures for power transformer testing. Also, it may be essential to reduce the effect of core saturation between the application of switching impulses.

7.6.2.2   Power-frequency withstand tests

In order to avoid saturation of the core during the 60 seconds overvoltage-induced tests, the test frequency may require increasing to more than twice the rated value. In this case the length of the application may be reduced as for power transformers.

7.6.2.3   Partial-discharge tests

For identical test conditions the allowable PD levels are similar to the current transformer values. Phase–phase values at $1.2U_m$ are 5 pC for liquid-immersed units and 20 pC for solid designs. The various quoted values are also valid for non-rated frequencies.

7.6.3   Capacitor voltage transformers

As a CVT consists of a voltage divider and an electromagnetic unit, the assessment of the insulation condition requires separation of the components for particular tests. The appropriate procedures are described in IEC/PAS 60044-5:2002 [S7/20]. Some of the more significant tests are summarized below.

7.6.3.1   Type tests

Monitoring of the insulation state is effected by capacitance and DDF measurements and, in the case of impulse tests, by the comparison of waveshapes before and after the test sequence. The type tests include the following.

- Wet power-frequency withstand-voltage tests on outdoor units (<300 kV system).
- Lightning impulse withstand-voltage tests comprising 15 consecutive applications of each polarity for a $U_m < 300$ kV system unit and three of each polarity for a $U_m \geq 300$ kV system unit. A standard wave is to be used but if loading on the generator is too large a front rise time of up to 8 μs may be applied.
- Two chopped impulse waves of negative polarity applied to both classes of CVTs. Full waves are required before and after the chopped waves.
- Wet switching impulse tests of 15 positive polarity waves applied to outdoor units for $U_m \geq 300$ kV.
- RIV tests if specified. Conditions as for inductive current voltage transformers. A PD test may be used as an alternative.
- Ferro-resonance tests if required.

7.6.3.2   Routine tests

As an insulation monitor capacitance and DDF measurements are made before and after the high-voltage tests.

- Dry power-frequency withstand-voltage test for one minute.
- Partial-discharge test to be applied. Alternative methods are allowable, including use of balanced circuit as in IEC 60270. Permitted levels for earthed neutral are 10 pC at U_m and 5 pC at $1.2\,U_m/\sqrt{3}$.
- Power-frequency withstand test is to be applied to the electromagnetic unit.

7.7   High-voltage power capacitors

The dielectric assessment of power capacitors ranges from tests on complete units of tens of microfarads for operation at approximately 7 kV (RMS) through to endurance tests (overvoltage cycling and ageing) on individual elements rated at, perhaps, 1.8 kV. Many of the capacitors are mounted outdoors in large banks for operation at the higher system voltages. They must be able to withstand adverse load-switching conditions and overvoltages and remain thermally stable under the very high internal electrical stresses. The testing of power capacitors requires dedicated supplies capable of providing several MVAR at test voltages of the order of 20 kV and for some cases to be effectively PD-free up to 10 kV. Tests for confirming the insulation performance are described in IEC 60871-1 and -2 [S7/21].

Type tests include thermal stability measurements, DDF determination at elevated temperature, an AC voltage withstand test and a lightning impulse-withstand test between terminals and container.

Routine tests are required for measurement of capacitance and DDF, for checking the AC withstand voltage and DC withstand voltage between terminals, and AC withstand voltage between terminals and container.

Special test. Endurance tests may be required on elements typical of the particular design and manufacturing process. Procedures and construction of elements for ageing and overvoltage cycling tests are specified in IEC 60871-2. Although partial-discharge measurements are not normally specified because of testing problems, facilities may be required to carry out such tests on complete capacitors. A PD-free source of up to, perhaps, 3MVAR and 15 kV (RMS) may be necessary. With the largest-rating capacitors, sensitivities of only 50 pC are achievable, which may not be considered significant by some authorities. A possible PD test sequence was described in AS 2897, 1986 [S7/21].

7.8   High-voltage rotating machines

The quality control of the insulating materials and insulation structures of large HV rotating machines during manufacture, assembly and commissioning is effected by application of a number of tests, some of which are listed below. The stator bars

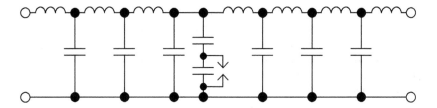

Figure 7.1   Equivalent circuit of PD defect in a power cable [5] [reproduced by permission of CIGRE]

- DDF measurements at operating and ambient temperatures are specified. For XLPE the expected values are $\leq 10 \times 10^{-4}$ and in the tens of mRs for EPR at operating voltages.
- Power-frequency withstand-voltage tests are required for long and short periods depending on the thermal conditions.
- Lightning-impulse withstand-voltage applications at temperatures of 95–100°C are followed by a 15-minute power-frequency withstand voltage. Switching impulses are specified for the heated system at voltages $\geq 300\,\text{kV}$.

7.9.2.2   Routine tests for extruded cables

- Power-frequency withstand test requirements range from five minutes at approximately $3.3 \times$ operating voltage to earth for the lower voltage systems to twice for 60 minutes at the highest voltages [S7/30].
- PD tests are specified at $1.5 \times$ operating voltage to earth. The method of calibrating when testing a 'long' cable using the conventional detection system (IEC 60270) is described in Reference S7/31.

The well-known circuit representation of a cable as a transmission line containing a PD defect is depicted in Figure 7.1 [5]. This reference presents a good analysis of the various developments in cable PD measurements. Figure 7.2 indicates how reflections are produced by a single discharge in a long cable. The example is for a partial discharge at 450 metres of an open-ended 480-metre-long cable [5]. The first pulse at the measurement end is followed by a reflection from the far end and a series of pulses set up as shown. The double pulses are due to the small 30-metre reflection distance at the far end. The intervals between the pulse pairs represent the transit times for the forward and reflected waves. Analysis of such data – probably in the presence of noise – enables the estimation of location of discharges and possible attenuation effects [5].

7.9.2.3   Installation tests for extruded cables

The commissioning of AC polymeric cables using HVDC for testing was not recommended by SC21(B1) of CIGRE (1990), since it is seen as being ineffective and possibly causing damage to the insulation, including the accessories assembled on site. AC tests of a 60-minute duration for voltage levels in the range of $2U_0$ at the

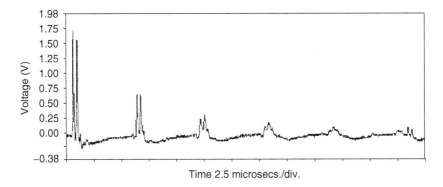

Figure 7.2   Attenuation of a multiple reflected PD pulse in a power cable [5] [reproduced by permission of CIGRE]

lower voltages down to $1.1U_0$ for 500 kV systems [S7/30] were subsequently proposed [CIGRE Brochure 173, Reference 6]. Twenty-four-hour tests at a voltage of U_0 may be employed if agreed, or higher if the supply is less than the power frequency. A test-frequency range of 20 Hz to 300 Hz is quoted in S7/30. Such power supply sources are now available, as reviewed in Chapter 6.

The special case of long extruded AC submarine cables (36–170 kV) is reported in *Electra* [7]. IEC revisions of the particular standards are related to the various proposals. A later CIGRE report [8] gives collation of practical test results obtained on installations using the levels recommended. The results confirm the conclusions of the earlier work.

7.10   Distribution and power transformers

The dielectric tests on distribution and, especially, power transformers are complex as the turn-to-turn, inter-winding and winding-to-earth insulations are required to be proved for two or more windings under steady-state and surge conditions [S7/32]. These requirements often necessitate the application of special arrangements – in design and when testing – in order to achieve the specified test levels.

7.10.1   Power-frequency overvoltage withstand tests

In three-phase units, particularly with three-limb cores, the single-phase induced power-frequency test can result in large voltage differences between phases but these may be reduced as indicated in Figure 7.3. A number of other configurations are possible [IEC 60076-3]. The recommended inter-phase air clearances are also given in IEC 60076-3.

For windings with one end nominally earthed (non-uniform insulation) it is usually impossible to induce the specified turn-to-turn voltage of twice normal and the correct HV terminal-to-earth voltage simultaneously, unless the neutral terminal is supplied

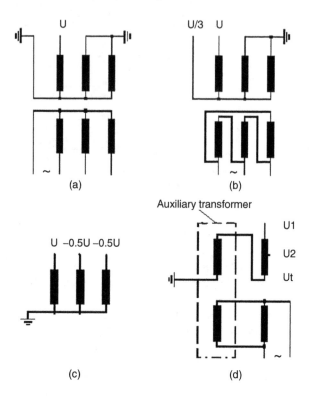

Figure 7.3   Single-phase induced-voltage tests to ground on three-phase and 1-phase transformers with non-uniform insulation [S7/32] [reproduced by permission of SAI GLOBAL]: (a) five-limb core (withstand overvoltage test and PD test); (b) three-limb core (withstand overvoltage test). Extra insulation at neutral; (c) three-limb core (withstand overvoltage test and PD test). 1.5U between phases; (d) 1-phase auto-transformer (withstand overvoltage test). Auxiliary testing transformer maintains correct voltage ratio between HV and Common terminals

from an independent source – for example, for an auto-transformer as in Figure 7.3(d). For such conditions the neutral end bushing and the transformer insulation levels may need to be of a higher value than required for normal tests.

Previously, the detection of failure during power-frequency short-term tests of 60 seconds – or less at higher frequencies to prevent saturation of the core during induced tests – was simply taken as collapse of the test voltage. If agreed, the Standard now includes a specification for partial-discharge measurements as a monitor of possible failure during the short-duration (ACSD) test sequence. The HV withstand-voltage test duration for frequencies greater than twice normal is given by the relationship 120 × (rated frequency/test frequency) seconds with a minimum of 15 seconds. In some cases the terminal voltage magnitude and waveform may be monitored via the bushing tap or an HV capacitor divider used for measuring the output voltage.

7.10.2 Partial-discharge tests

If a partial-discharge test is specified, the conventional technique applied is that described in Chapter 6 utilizing the HV bushing (and tap) as the coupling capacitor for the higher-voltage units. For lower-voltage units, e.g. dry-type distribution transformers, separate coupling capacitors are required [S7/32]. The PD test is usually carried out as a single-phase test. The test sequences for both the short-term duration (ACSD) and the long-term duration (ACLD) tests are similar. The duration of each and the voltage magnitudes are given in the Standard. The ACLD test is considered to be a quality-control test and not for proving the design.

In both types of test a five-minute 'conditioning' period at $1.5U_m/\sqrt{3}$ kV to ground is induced before increasing the value to the withstand test voltage for the ACSD test and to $1.7U_m/\sqrt{3}$ for the ACLD test. After lowering the voltage to $1.5U_m/\sqrt{3}$ kV, partial discharges are measured continuously for assessment during a five-minute period for the ACSD condition and during a sixty-minute period ($U_m > 300$ kV)/30-minute period ($U_m < 300$ kV) for the ACLD test. Also, values in the final five-minute periods at $1.1\ U_m/\sqrt{3}$ are included in the assessment. Other PDs measured throughout the sequences are not considered significant in judging the success of the tests.

Single-phase tests on three-phase transformers are preferably made with connections as in Figure 7.3 (c). The maximum voltage is applied for the one-minute equivalent as determined from the test frequency. The other times are independent of frequency. Allowance must be made for synchronizing the discharge detector with the test frequency to assist with interpretation of the PDs.

7.10.2.1 Interpretation of PD measurements

The introduction of PD measurements during the short-duration over-potential test (ACSD) may enable the detection of the inception and extinction voltages and the discharge magnitudes to be determined if a disturbance is present during the application. Due to the rapid rise of voltage it may be difficult to estimate the precise time and values. An automated detection system would be of assistance. Because of the variabilities this part of the test is not used for PD assessment purposes. However, the data obtained might be of assistance to designers and the manufacturer if a fault is suspected.

In both types of test a continuous PD value of >300 pC after the high-voltage application is not acceptable. The value at the conclusion is to be <100 pC with the voltage 10 per cent above operating voltage. This is based on a background interference of 100 pC that might be considered to be a high value. In some specifications the allowable value is 50 pC at $1.2 \times$ maximum phase operating voltage to ground. It is also noted that bursts of apparent discharges higher than 300 pC may be ignored. Research has shown that such unsustainable discharges can be produced by some types of fault and, perhaps, should be recorded for future reference. The acceptance or otherwise of PD values near the limit can be difficult as the error in measurement may be as much as 20 per cent, depending on the conditions and calibration.

proportionate magnitude levels) in the recorded waveform at reduced voltage and at full test voltage sometimes necessitates careful consideration and, perhaps, agreement in interpretation between the purchaser and manufacturer. This may be particularly relevant for chopped-wave tests where sudden changes occur at the time of chopping (effected after a few μs) tending to mask any high-frequency changes due to a fault. Depending on the number of recording channels available, and the particular case, time bases chosen are in the range 0–10/25 μs for the earlier part of the wave and 0–100 μs for the later times.

The detection of failure during switching-impulse tests is simpler than for lightning-impulse tests, as the voltage distribution within the windings and phases is effectively linear and similar to power-frequency conditions. Records are made of the line voltage and, probably, of the neutral current waveshapes. The required waveform must have a virtual rise time of at least 100 μs with a time greater than 90 per cent of the test value for 200 μs and a time from virtual zero to the first zero crossing of $\geq$ 500 μs. The recording time bases required are from 100/500 μs to 1 000/5 000 μs. Typical examples are given in IEC 60076-4. The front must be such as to give an approximately uniform distribution within the winding.

A major factor when applying switching surges is the possibility of core saturation, resulting in waveform distortion. In order to allow comparison between records at different levels it may be necessary to apply lower-level impulses of opposite polarity, or even direct voltage, to establish acceptable core conditions. The withstand voltage may be applied directly to the winding under test or induced from a lower-voltage winding. The voltages are approximately in the ratio of the turns.

During switching surge tests on three-phase transformers it is important to ensure the clearances between phases are adequate [IEC 60076-3]. Tests are usually carried out with negative impulses, as air flashover voltages are greater than for positive waves. The non-restoring internal insulation is assumed to have the same strength for both polarities. Superimposed oscillations due to inter-phase capacitive coupling may produce higher than expected voltages [IEC 60076-4], which must be controlled, possibly by resistive damping.

The test is successful if no sudden change occurs in the shape of the line voltage. Corresponding changes in the neutral current, if recorded, may also indicate a problem.

7.10.3   Summary of transformer HV test requirements

The appropriate high-voltage tests for voltage (kV) classes $U_m \leq 72.5$, $72.5 < U_m \leq 170$, $170 < U_m < 300$ and $U_m \geq 300$ are tabulated in IEC 60076-3 [S7/32]. The tests relate to uniform and non-uniform insulated transformers. The appropriate withstand-voltage magnitudes ('one-minute' test) and impulse levels for each operating system are given in the Standard Tables.

- **Switching-impulse withstand voltage (SI)**: If the ACSD test is not applied the SI test is specified for those units subjected to the ACLD test.
- **Lightning-impulse withstand voltage (LI)**: When specified this test covers chopped waves (see Figure 7.6) as a special test.

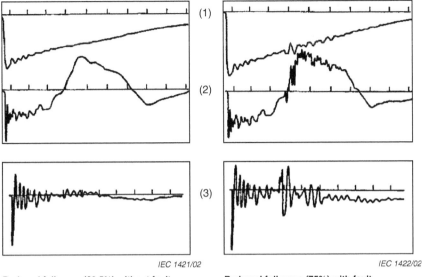

IEC 1421/02   IEC 1422/02

Reduced full wave (62.5%) without fault   **Reduced full wave (75%) with fault**

(Amplitudes not equalized)
1 Applied impulse, 100 μs sweep
2 Capacitively transferred current from the shorted adjacent winding to earth, 100 μs sweep
3 Neutral current, 100 μs sweep
Note: Failure indicated after 30 μs in voltage, capacitively transferred current and neutral current oscillograms

Figure 7.5   Lightning-impulse full wave failure – interlayer breakdown in coarse-step tapping winding of a 400/220 kV transformer [reproduced by permission of SAI GLOBAL]

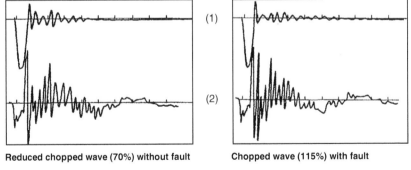

Reduced chopped wave (70%) without fault   **Chopped wave (115%) with fault**

1 Applied impulse, chopped wave, 50 μs sweep
2 Capacitively transferred current from the shorted adjacent winding to earth, 50 μs sweep
Note: Failure indicated immediately after chopping in both the voltage and capacitively transferred current oscillograms

Figure 7.6   Lightning-impulse chopped-wave failure – between turns breakdown in fine-step tapping winding of a 400/220 kV transformer [reproduced by permission of SAI GLOBAL]

- **Separate-source AC withstand voltage**: The test checks the insulation from the neutral to ground for non-uniform windings and the whole of the insulation to ground for uniform windings.
- **Short-duration induced AC withstand voltage (ACSD)**: A PD test may be applied if agreed between manufacturer and user.
- **Long-duration induced AC voltage (ACLD)**: The test includes PD measurements for 30/60 minutes at overvoltages to confirm the absence of damaging PDs during service.

In cases where lightning-impulse chopped waves are specified a possible sequence might be 1FW (50–75 per cent), 1FW (100 per cent), several CW for reference (50–75 per cent), 2CW (100 per cent), 2FW (100 per cent), 1FW (50–75 per cent). The chopping time should be in the range 2–6 μs. Failure is determined by careful comparison of records taken during the reduced chopped waves and those at the full level (Figure 7.6). It is assumed that the chopping times at reduced and full voltage are approximately equal. A slight difference in chopping times (perhaps $<0.1\,\mu$s) might result in changes similar to those caused by a fault during application of a CW.

7.10.4   Additional tests

In addition to the well-established tests reviewed above, a number of newer tests are being trialled especially for in-service monitoring. These include more advanced PD systems, DC charging and relaxation measurements and frequency response analysis (FRA). Details of some of the various techniques are considered in Chapters 6, 8 and 9.

7.11   Dielectric testing of HVDC equipment

The choice of lightning and switching impulses for HVDC equipment depends on the insulation coordination levels of the system. The relevant direct voltage tests as related to the continuous DC operating voltages are, in general, agreed between manufacturer and user. A discussion of voltage stresses and test requirements for converter stations and HVDC cables is presented in References 9 and 10. Guidance on coordination procedures is considered in IEC 60071-5 [S7/34] for converter stations.

An important aspect of the protective methods is the correct application of surge arresters, especially related to the thyristor valves. The testing of such valves is specified in IEC 60700-1 [S7/35], including methods for determining the test levels based on given test safety factors. For example these are, respectively, 1.6 and 1.3 during the one-minute and three-hour DC tests of external insulation to earth. The three-hour value is reduced to 0.8 for a valve test. Partial-discharge tests are required during these applications, the PD values and repetition rates being recorded during the last hour. The number of pulses greater than 300 pC must be $\leq$ 15/minute – a spread of allowable values of up to 2 000 pC (1/minute) is allowed. The measurements are according to IEC 60270 and must incorporate methods for counting and recording the pulses. AC partial-discharge tests are also required, the limit being 200 pC during a 30-minute test at a voltage based on the equivalent peak-to-peak, steady-state operating level and a safety factor of 1.15. The standardized lightning and switching surge (three of each polarity) test levels are 1.15 × surge protective level of the valve arrester and 1.2 for steep surges. The various complexities in testing the dielectric strength of thyristor valves when subjected to the combined effects of DC, AC and surge conditions are detailed in the standard.

Recommendations for tests on paper-insulated cables for DC transmission voltages up to 800 kV are included in a CIGRE report [11]. It is recommended that the factory high-voltage acceptance test be a negative DC voltage of 1.8 × U_0 applied for 15 minutes. The allowable power factors of the manufactured cable lengths are specified – for example, for an oil-filled cable rated at 400 kV DC, the value should not exceed 46 × 10^{-4} at a maximum RMS stress of 20 kV/mm. Type tests include combined polarity reversal/load cycles with a voltage of 1.4 × U_0. Impulse- (lightning- and switching-) superposed tests with the cable at a negative DC voltage of U_0 are described for each case. Positive and negative impulses are to be applied with recommended minimum values of 1.15 × lightning impulse protective level and 1.15 × switching surge level of the particular installation. Details of the procedures are given in the report. The suggested site test is a negative DC voltage of 1.4 U_0 applied for 15 minutes.

Extruded insulation for DC transmission cables has been limited to commercial applications up to voltages of 80 kV and 150 kV as in the HVDC Light VSC projects in Australia (2000/2002) including extruded cable lengths up to 180 km. A few failures have occurred in other schemes, possibly due to space-charge problems following polarity reversals. Material developments and the introduction of the new converter technology (voltage source converter, VSC), which does not reverse the polarity, may give impetus to the wider use of extruded HVDC power cables [12]. This is reflected in the proposals for establishing test procedures for such cables in the range 80–250 kV included in the report by CIGRE WG 21.01 [13]. A number of the proposals are similar to those in Reference 11, although allowance is made, for example, for the longer time constants. The type load tests at 1.85 × U_0 include 24 + 24 hours load cycles and differences in impulse tests for a VSC system. AC tests are recommended where practical for monitoring insulation quality. The factory-acceptance and post-installation tests are to be HVDC.

Bushings for the converter transformers and reactors must be tested separately under HVDC conditions, as the designs are not similar to AC units, the voltage stresses being governed by the resistivities and temperatures of the insulating materials. Long-term PD tests would probably be required.

The testing of the converter transformers requires application of HVDC voltages to the valve windings in addition to appropriate AC and impulse tests to all windings. The procedures aim to cover the condition of superimposed AC and DC voltages. These include PD tests under DC conditions requiring long-term testing to allow the correct voltage distributions to be established. After 50 years of testing experience in the industry, IEC has published a standard [IEC 61378-2, S7/36]. A Trial Use Test Code was published by the IEEE in 1999 [C57.129]. The series reactors are subjected

S7/8 IEC 60437: Radio interference test on high-voltage insulators (AS IEC 60437–2005)
S7/9 IEC 60099: Surge arresters:

Part 1: Non-linear resistor type gapped arresters for AC systems (AS 1307.1)
Part 4: Metal-oxide surge arresters without gaps for AC systems

S7/10 AS 1307.2: Surge arresters – Part 2 Metal-oxide arresters without gaps for AC systems
S7/11 IEC 60694 (Ed. 2.2, 2002): Common specifications for high-voltage switchgear and controlgear standards
S7/12 IEC 62271-100 (Ed. 1.1, 2003): High-voltage switchgear and controlgear – Part 100: High-voltage alternating-current circuit-breakers (AS 62271.100, 2005)
S7/13 IEC 62271-102 (Ed.1.0, 2003): High-voltage switchgear and controlgear – Alternating current disconnectors and earthing switches (AS 62271.102, 2005)
S7/14 IEC 62271-200 (Ed.1.0, 2003): High-voltage switchgear and controlgear – Part 200: AC metal enclosed switchgear and controlgear for rated voltages above 1 kV and up to and including 52 kV (AS 62271 200, 2005)
S7/15 IEC 62271-201 (Ed.1.0, 2006): High-voltage switchgear and controlgear – Part 201: AC insulation-enclosed switchgear and controlgear for rated voltages above 1 kV and up to and including 52 kV
S7/16 IEC 62271-203 (Ed. 1.0, 2003): High-voltage switchgear and controlgear – Part 203: Gas-insulated metal-enclosed switchgear for rated voltages above 52 kV (AS 62271.203, 2005)
S7/17 IEC 60214-1 (Ed.1.0): Tap-changers –Part 1: Performance requirements and test methods (AS 60214.1, 2005)
S7/18 IEC 60137 (Ed. 5.0, 2003): Insulated bushings for alternating voltages above 1 000 V (AS 1265, 1990)
S7/19 IEC 61464 TR2 (Ed. 1.0, 2003): Insulated bushings – Guide for the interpretation of DGA in bushings where oil is the impregnating medium of the main insulation (generally paper) (AS TR 61464, 2006)
S7/20 IEC 60044: Instrument transformers (AS 60044 Parts 1, 2, 5)
60044-1 (Ed. 1.2, 2003) – Part 1: Current transformers
60044-2 (Ed. 1.2, 2003) – Part 2: Inductive voltage transformers
60044-5 (Ed. 1.0, 2004) – Part 5: Capacitor voltage transformers
S7/21 IEC 60871: Shunt capacitors for AC power systems having a rated voltage above 1 000 V (AS 2897, 1986, withdrawn)
60871-1 (Ed. 3.0, 2005) – Part 1: General
60871-2 Ed. (2.0, 1999) – Part 2: Endurance testing
S7/22 IEC 60034: Rotating electrical machines
60034-1 (Ed. 11.0, 2004) – Part 1: Rating and performance (AS 1359.101, 1997)
60034-15 (Ed. 2.0, 1995) – Part 15: Impulse voltage withstand levels of rotating AC machines with form-wound stator coils
S7/23 IEEE Std. 286, 2000, IEEE: Recommended practice for the measurement of power factor tip-up of electric machinery stator coil insulation

S7/24 IEEE Std. 95, 2002, IEEE: Recommended practice for insulation testing of AC electric machinery (2 300 V and above) with high direct voltage

S7/25 IEEE Std. 1434, 2000, IEEE: Guide to measurement of partial discharges in rotating machinery

S7/26 IEC 60055-1 (Ed. 5.1, 2005): Paper-insulated metal-sheathed cables for rated voltages up to 18/30 kV (with copper or aluminium conductors and excluding gas-pressure and oil-filled cables) – Part 1: Tests on cables and their accessories (AS 1026, 2004)

S7/27 IEC 60141: Tests on oil-filled and gas-pressure cables and their accessories 60141-1 (Ed. 3.0, 1993) – Part 1: Oil-filled, paper or polypropylene paper laminate insulated, metal-sheathed cables and accessories for alternating voltages up to and including 500 kV

60141-2 (Ed. 1.0, 1963/67) – Part 2: Internal gas-pressure cables and accessories for alternating voltages up to 275 kV

60141-4 (Ed.1.0, 1980/90) – Part 4: Oil-impregnated paper-insulated high pressure oil-filled pipe-type cables and accessories for alternating voltages up to and including 400 kV

S7/28 IEC 60502: Power cables with extruded insulation and their accessories for rated voltages from 1 kV ($U_m = 1.2$ kV) up to 30 kV ($U_m = 36$ kV) (AS 1429.1, 2006)

60502-2 (Ed. 2.0, 2005) – Part 2: Cables for rated voltages from 6 kV ($U_m = 7.2$ kV) up to 30 kV ($U_m = 36$ kV)

60502-4 (Ed. 2.0, 2005) – Part 4: Test requirements on accessories for cables with rated voltages from 6 kV ($U_m = 7.2$) up to 30 kV ($U_m = 36$ kV)

S7/29 IEC 60840 (Ed. 3.0, 2004): Power cables with extruded insulation and their accessories for rated voltages above 30 kV ($U_m = 36$ kV) up to 150 kV ($U_m = 170$ kV) – Test methods and requirements (AS 1429.2, 1998)

S7/30 IEC 62067 (Ed. 1.1, 2006): Power cables with extruded insulation and their accessories for rated voltages above 150 kV ($U_m = 170$ kV) up to 500 kV ($U_m = 550$ kV) – Test methods and requirements

S7/31 IEC 60885-3 (Ed. 1.0, 1988): Electrical test methods for electric cables – Part 3: Test methods for partial-discharge measurements on lengths of extruded power cables (AS 1660.3, 1998)

S7/32 IEC 60076-1 (Ed. 2.1, 2000): Power transformers – Part 1: General (AS 60076 Parts 1, 3, 4, 11)

60076-3 (Ed. 2.0, 2000): Power transformers – Part 3: Insulation levels, dielectric tests and external clearances in air

60076-4 (Ed. 1.0, 2002): Power transformers – Part 4: Guide to lightning impulse and switching impulse testing – Power transformers and reactors

60076-11 (Ed. 1.0, 2004): Power transformers – Part 11: Dry-type transformers

S7/33 IEC 61181 (Ed. 1.0, 1993): Impregnated insulating materials – Application of dissolved-gas analysis (DGA) to factory tests on electrical equipment

S7/34 IEC/TS 60071-5 (Ed. 1.0, 2002): Insulation coordination – Part 5: Procedures for high-voltage direct-current (HVDC) converter stations

S7/35 IEC 60700-1 (Ed. 1.1, 2003): Thyristor valves for high-voltage direct-current (HVDC) power transmission – Part 1: Electrical testing

S7/36 IEC 61378-2 (Ed. 1.0 (2001): Converter transformers – Part 2: Transformers for HVDC applications

S7/37 IEC 61325 (Ed. 1.0, 1995): Insulators for overhead lines with a nominal voltage above 1 000 V – Ceramic or glass insulator units for DC systems – Definitions, test methods and acceptance criteria

S7/38 IEC 60855 (Ed. 1.0, 1985/1999): Insulating foam-filled tubes and solid rods for live working

S7/39 IEC 60900 (Ed. 2.0, 2004): Live working – Hand tools for use up to 1 000 V AC and 1 500 V DC

S7/40 IEC 60903 (Ed. 2.0, 2002): Live working – Gloves of insulating material

7.16   Problems

1. Analyse and discuss the function and influence of Standards on the application of insulation in the high-voltage power industry. Highlight their importance economically and technically and the need to standardize assessment/condition monitoring methods for equipment and materials whenever appropriate.

2. Determine the PD pulse separation at the measurement end of the cable in Section 7.9.2 for a source located at 100 metres from the far end. Why is the apparent pulse velocity less than the value based on a cable-relative permittivity of 2.2?

parameters, and can be incorporated within the equipment itself without being intrusive. The parameters that can be monitored range from chemical species (degradation by-products, contaminants and wear debris), internal/external strain fields, pressure, fracturing and fatigue to vibrations, acoustic emission and temperature.

The high expectations of optical fibres as information carriers in communication systems has been justified by their performance over the past three decades. Due to the high bandwidth and low attenuation properties, each fibre is capable of replacing 1 000 copper wires. As a sensor, an optical fibre can carry much more information. Light that is launched into and confined to the fibre core propagates along the fibre unperturbed unless experiencing an external influence. Any disturbance to the fibre alters the characteristics of the guided light, which can then be monitored and related to the magnitude and location of the disturbing influence. The characteristics of the light that may be monitored in sensing applications include amplitude, phase, wavelength, polarization, modal distribution and time-of-flight.

8.2.1 Basic physics of optical-fibre sensing

There are many different types of optical-fibre sensor. According to the method by which the light is modulated in the fibre, they fall into four main groups [8], as follows.

- **Intensiometric**: Intensiometric optical-fibre sensors are based on the modulation of light intensity in the fibre and are generally configured as a distributed sensor. The light intensity may change in the form of fracture loss, time-of-flight, refractive index, amplitude and wavelength.
- **Interferometric**: Optical-fibre interferometers are generally intrinsic sensors. With the light from a source equally divided and directed through two (or more) fibre-guided paths, the beams are then recombined to mix coherently. Fingerprint types can be established in relation to the optical phase difference experienced by the two light beams. Single-mode fibres are used and very high sensitivity can be achieved.
- **Polarimetric**: The principle of polarization sensing is based on the birefringence within the fibre. When launching a polarized light beam along the principal axis of a single-mode fibre at 45°, the two orthogonal polarization eigenmodes are equally excited and degenerated. The degeneracy is increased if the fibre is subjected to an external force and a phase difference is induced in the two eigenmodes. The state of polarization in the fibre can then be related to the magnitude of the external force. These types of sensor are not as sensitive as interferometers, but have longer sensing lengths.
- **Modalmetric**: Optical fibres are microphonic and the detection sensitivity is based on the modulation in the distribution of modal energy in the fibre. This type of sensor is sensitive, but the modulation of the modal pattern is normally nonlinear in relation to all disturbances, resulting in fading and drifting in the output signal. Usually, these sensors are used in conjunction with other techniques such as microbending and polarimetry.

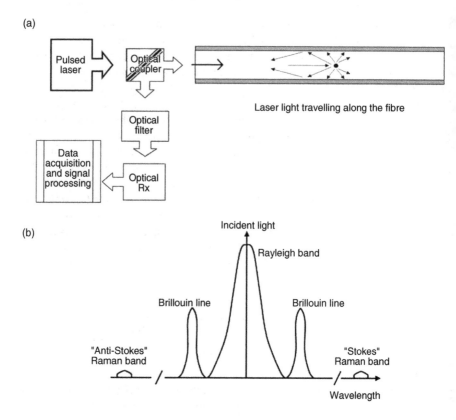

Figure 8.4   The block diagram of an interferometer optical-fibre sensor and the back-scatter spectrum [8,10]

The block diagram of an interferometer optical-fibre sensor is shown in Figure 8.4. Optical-fibre sensors are rapidly proving to be a vital new tool for insulation researchers. The devices have already found uses in temperature, pressure and vibration measurements. Speciality fibre technology is opening up more opportunities in insulation condition monitoring such as PD measurements and transformer oil moisture detection. A few typical applications are listed below, although further investigations are necessary before they can be fully utilized in the industry.

8.2.2   Optical-fibre PD sensors

8.2.2.1   Partial discharge measurements on generators and GIS

The acoustic emissions from partial discharges can be detected by optical-fibre sensors. Researchers have investigated different types of such sensors and have found that achieving sufficient sensitivity is the most difficult problem. In a GIS chamber, the typical acoustic vibration generated by PDs is 1×10^{-5} g to 1×10^{-2} g, where g $= 9.81$ m/s^2. Most of the acoustic energy from PDs is at ultrasonic frequencies,

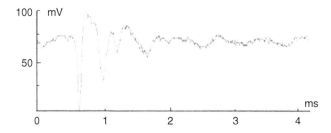

Figure 8.5   *The typical responses of an optical-fibre sensor to PDs in a genera-*
tor stator bar tested in the laboratory. The discharge magnitude was
approximately 500 pC [8]

i.e. around 10 kHz. At this frequency, the acoustic vibration acceleration can be
translated into displacements in the order of 10^{-14} m to 10^{-12} m. For generators, the
magnitude of PDs is much higher than in GIS, but even so more sensitivity is required
before a reliable sensor can be used. The PD energy detected depends on the length
of the sensing fibre, the acoustic power and the numerical aperture of the fibre.

Monash University and FFT Ltd have jointly developed the Foptic μStrain sensor
and associated instrument [8]. The technique is based on a well-known fibre-optic
interferometric principle that can achieve a wide-frequency bandwidth of 1 MHz.
The sensor is less complex and of lower cost than other optical-fibre sensing systems.
Only a short sensing length of 6 mm is necessary, which makes the installation easier.
The sensor has been tested on a generator stator bar in the laboratory and the typical
response of this sensor to PDs in the generator stator bar is shown in Figure 8.5.

8.2.2.2   Partial discharge measurements on transformers

PD measurements on transformers using electrical methods are very difficult because
of the low level of discharge to be measured and the large interference at site. In a
330 kV substation, corona discharges from the overhead transmission lines can reach
3 000 pC or more, while the PD level to be measured could be less than 100 pC. In
such a noisy environment to detect PDs smaller than 5 per cent of noise would be
impossible by conventional methods. Optical-fibre sensors are immune from electrical interference and can be installed around the winding, which is excellent for PD
monitoring on transformers. However, the sensitivity of optical-fibre sensors is not
sufficiently high for such an application.

A novel optical sensor was developed [9] that uses an optical-fibre head instead
of a piezo-electric sensor for the detection of acoustic emission from the PD source,
as shown in Figure 8.6. The acoustic pressure changes the shape of the optical fibre,
which carries laser light injected from terminal 1. As a consequence, the phase of the
light will change according to the pressure. When reflected from the mirror at the
fibre head 3, the change will be detected and converted to an electrical signal. The
magnitude of the signal will vary according to the acoustic wave pressure from the
PD source.

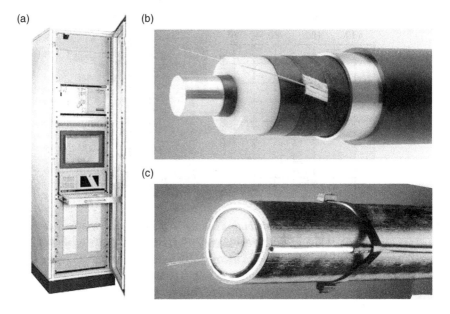

Figure 8.9   The DTS system made by Sensa Ltd and the cables with fibre sensors integrated (b) or laid on the surface (c) [10] [courtesy of Sensa Ltd, http://www.sensa.org]

- The temperature record provides important information about the thermal history of a cable that can then be used to analyse insulation ageing for life prediction and maintenance planning.
- It assists in system operation by generating an alarm when the temperature exceeds the limit.
- The real-time temperature measured is important for cable emergency load management [12].

Figure 8.9 shows the DTS system widely used in underground HV power cables [12]. The optical fibre can be integrated inside the cable (under the sheath), attached to the surface or in the duct next to the cable.

There are various optical-fibre techniques for temperature sensing. Figure 8.10 shows the sensing system using optical time domain reflectometry. The laser light is injected through the optical coupler to the optical-fibre sensor and reflected back at the mirrored end. While travelling along the fibre, the light is distorted by the temperature in a certain pattern. The reflected light is then received after coupling back to the detector and compared with the injected light in its phase and magnitude, which can then determine the temperature at different locations along the fibre:

- emission-to-sampling time defines the measurement position;
- the light intensity at a particular location gives the temperature.

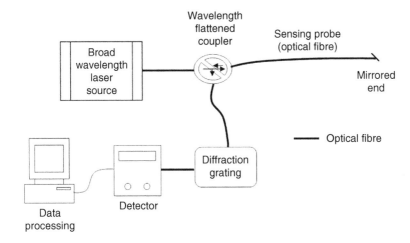

Figure 8.10   The block diagram of a distributed optical fibre sensing system [8]

An important characteristic of DTS is the temperature sensitivity, which can be increased by the use of Raman scattered light.

Point sensors are also available to detect temperature at important parts of a cable, e.g. cable joints and terminations. The data can be collected locally and transmitted to the control centre or laboratory for continuous monitoring and analysis.

8.2.4   Advantages and disadvantages of optical-fibre sensors

The principal advantages of optical fibres over conventional sensors are:

- small size, low weight and robustness;
- low unit cost for sensor;
- corrosion resistance, high-tensile strength and high fatigue life of optical fibre;
- non-conductivity and immunity to electromagnetic interference;
- very wide-frequency bandwidth and fast response times;
- ability to measure very wide temperature range;
- high spatial resolution and high sensitivity for some events;
- simultaneous sensing of more than one parameter.

The major disadvantages of optical-fibre sensors are that:

- the optical fibre may be sensitive to different influences, which could require the isolation of unwanted parameters;
- the detection and signal-processing instruments can be complicated and costly;
- the long-term stability of an optical-fibre sensor in particular power equipment environments needs examination;
- it has a low sensitivity to partial discharge at this stage of development.

8.3   Directional sensors for PD measurements

For HV equipment with earthed metal sheaths or enclosures such as transformers, power cables and HV motors, electrical noise sources are mainly external – for example, corona on the conductor connected to the HV terminal. On such a conductor, PD pulses travel from the equipment terminal outwards and electrical noise travels inwards, as shown in Figure 8.11 for a shielded power cable. The propagation direction of the pulses is therefore a useful indicator for identifying the noise. The conventional method for determining the direction of a travelling wave on a conductor is to connect two HV capacitors to the conductor at a certain distance apart. External noise will arrive at the outer sensor first and at the other after a short time delay. The distance has to be >3 m to make the delay sufficiently long for the electronic circuit to make the comparison. The use of HV capacitors for this purpose is expensive and can cause additional risks to the HV network.

8.3.1   Directional coupler sensor

EHV underground power cables are critical for the reliability of electrical transmission systems. It has been found that cable accessories such as joints and terminals are the weakest parts in terms of the insulation strength. For the voltage above 220 kV, it is required that the PD level should be very low or undetectable under the operating voltage. The directional coupler system (LDS-5, Lemke Diagnostics AG) shown in Figure 8.12 provides an effective means to discriminate external interference and corona noise from the testing voltage supply.

The directional couplers are installed between the cable sheath and the semiconductive layer. There are four outputs from the couplers, which are marked A, B, C and D in Figure 8.12. The noise travelling from left to right will be picked up by A and C outputs and from right to left picked up by B and D. However, PDs from the

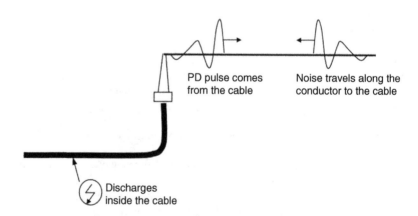

Figure 8.11   *The PD pulse from a cable and the external noise travelling in opposite directions*

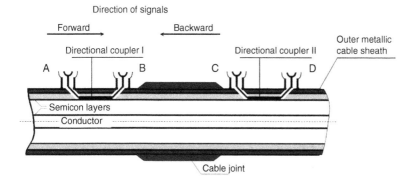

Figure 8.12   The directional couplers installed on both sides of a cable joint for PD detection at the high measurement sensitivity [courtesy of Lemke Diagnostics AG, URL: http://www.ldic.ch, http://www.hvdiagnostics.de]

cable joint will be picked up only by B and C, which are distinguished from the noise originated on both sides.

8.3.2   Directional field sensor

The new directional sensor developed jointly by Monash University and Insultest Australia Ltd does not require a connection to the HV conductor [13]. It consists of a 1.2-metre-long rectangular detector that picks up the electrical and magnetic fields associated with the PDs. The fields reflect the voltage and current of a partial discharge. With the correct configuration, the voltage and current signals produced by the PDs from the HV equipment will have the same polarity, but those due to external noise will be of opposite polarities. By comparing the polarities of the two signals, the direction of travelling pulses can be detected. The principle of the system is shown in Figure 8.13.

Because of its high sensitivity the sensor can be installed at ground level, with a distance of 200–300 mm from the HV conductor (busbar), as shown in Figure 8.14. There is no need to connect the sensor to the HV conductor. The sensors have been installed at two power stations giving some encouraging results. A ring-type directional sensor was also developed, which can easily be installed around the cable terminal or transformer bushing with a clearance of 100–300 mm. Investigations are being carried out to improve further the measurement sensitivity for the detection of PDs in transformers and XLPE cables [14,15].

Outputs of the directional sensor of Figure 8.14(b) for PDs travelling along the busbar in different directions are given in Figure 8.15. It can be seen that for a PD travelling from left to right, the first peaks of the electric and magnetic probe outputs have the same polarity, whereas, for PDs travelling from right to left, the polarities are opposite.

4. Sedding, H.G., Campbell, S.R., Stone, G.C., and Klempner, G.S., 'A new sensor for detecting partial discharge in operating turbine generators', *IEEE Transactions on Energy Conversion*, December 1999;**6**(4):700–6

5. Judd, M.D., Yang, L., and Hunter, I.B.B., 'Partial discharge monitoring for power transformers using UHF sensors, Part 1: Sensors and signal interpretation', *IEEE Electrical Insulation Magazine*, May/June 2005;**21**(2): 5–14

6. Judd, M.D., Yang, L., and Hunter, I.B.B., 'Partial discharge monitoring for power transformers using UHF sensors, Part 2: Field experience', *IEEE Electrical Insulation Magazine*, May/June 2005;**21**(3):5–13

7. Judd, M.D., Farish, O., Pearson, J.S., and Hampton B.F., 'Dielectric windows for UHF partial discharge detection', *IEEE Transactions on Dielectrics and Electrical Insulation* December 2001;**8**(6):953–8.

8. Tapanes, E., Oanca, I., Katsifolis, J., and Su, Q., 'The innovative use of optical fibres for condition monitoring of high voltage equipment', *Proceedings of the 5th International Conference on Optimization of Electric and Electronic Equipment*, Brasov, Romania, 15–17 May 1996, pp.1–24

9. Blackburn, T.R., Phung, B.T., James, R.E., 'Optical fibre sensor for partial discharge detection and location in high-voltage power transformer', *Proceedings of the 6th Annual Conference on Dielectric Materials, Measurements and Applications*, 7–10 September 1992, pp. 33–6

10. Nokes, Geoff, 'Optimising power transmission and distribution networks using optical fibre distributed temperature sensing systems', *Power Engineering Journal*, December 1999:291–6

11. Su, Q., Li, H.J., Tan, K.C., 'Hotspot location and mitigation for underground power cables', *Proceedings of the IEE on Generation, Transmission and Distribution*, November 2005;**152**(6):934–8

12. Li, H.J., Tan, K.C., and Su, Q., 'Assessment of underground cable ratings based on distributed temperature sensing', *IEEE Transactions on Power Delivery*, October 2006;**21**(4):1763–9

13. Su, Q., 'Research and development on insulation condition monitoring in Australia', keynote speech at the 1st International Conference on Insulation Condition Monitoring of Electrical Plant, Wuhan, China, 24–26 September 2000, pp. 19–24

14. Su, Q., 'Development of a directional sensor for noise discrimination in partial discharge measurements', *Proceedings of IEEE conference on Precision Electromagnetic Measurements*, Sydney, May 2000, pp.1–4

15. Su, Q., and Sack, K., 'Non-contact directional sensors for PD measurements', Techcom 2003, Sydney, May 2003, pp. 26–31

16. Australian Standard AS 1026:1992, Electrical Cables – Impregnated Paper Insulated, for working voltages up to and including 33kV

17. Australian/New Zealand Standard AS/NZS 1429.1:2000 and 1429.2:1998, Electrical Cables – Polymeric Insulated

8.6   Problems

1. What are the advantages of optical sensors for PD measurement on HV equipment? Why are they not widely used in industry?

2. The bandwidth of a detector can significantly affect PD measurement sensitivity and the effectiveness of noise immunity. Give three examples for each case of applications of narrow, wide and UHF band detectors in PD measurements respectively.

3. Why does the directional field sensor discussed in this chapter detect the direction of travelling PD signals along a conductor? What are the principles of electric and magnetic field measurements?

4. The direction of a travelling PD pulse along a conductor can also be determined by placing two capacitive coupling sensors at a certain distance apart along the conductor. What is the principle of this directional sensor? What are the limitations of this type of directional sensor (in frequency band, distance between sensors etc.)?

Chapter 9

Online insulation condition monitoring techniques

- Noise-mitigation techniques
- Non-electrical online condition monitoring techniques
- Electrical online condition monitoring techniques

Offline condition assessment of the insulation in HV equipment is applied extensively in order to minimize the possibility of failure in service. However, the required testing and measurement procedures are sometimes impractical, costly and not indicative of operating conditions. This chapter considers the alternative of online monitoring by means of which more continuous assessment is possible under operating conditions. Some of the developments during the past decade are discussed, including a number of new techniques aimed at overcoming many difficulties in implementing in-service measurements. The cost, reliability and convenience of the new systems need to be balanced against the savings effected by reduction in outages and the extension of life of the insulating materials.

9.1 The main problems with offline condition monitoring

Although offline insulation tests, either destructive or non-destructive, are valuable for the assessment of HV equipment condition, there exist several disadvantages.

1. The equipment has to be taken out of service, which may cause an unnecessary outage or reduction in electricity supply. For example, a large steam turbine generator has to run continuously during the maintenance interval of 3–4 years. A power transformer at the critical location in a power system may need 3–6 months' preparation for loads to be transferred before it can be taken out of service.
2. The equipment cannot be continuously monitored during operation. A fault may occur between planned offline tests. Such periodic measurements cannot guarantee to detect all developing defects of significance.

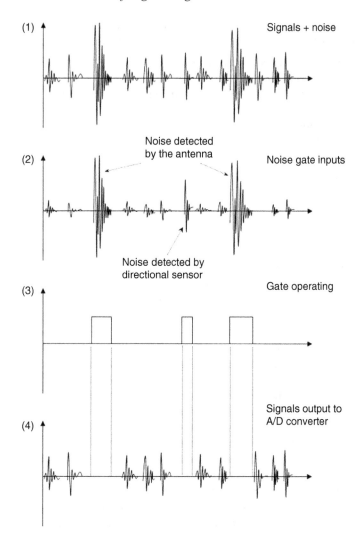

Figure 9.2   Noise gating process: (1) signal + noise; (2) noise detected by various sensors; (3) gating windows generated by noise; (4) noise removed by gate leaving PD signals to be measured

noise. Noise entering the HV terminals may also be detected by a directional sensor, as discussed in Chapter 8.

Noise may also be removed in the software of the measurement system. The detected noise can be registered in the computer program by a triggering channel and subsequently removed. For example, in the peak detection and A/D converter circuit, if the noise input channel shows a positive sign or the noise exceeds the trigger level, an additional bit will be attached to the digital signal from the A/D converter and later

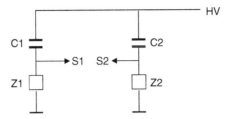

Figure 9.3   PD detection using the balanced circuit

removed in the analysis program. Software noise gating is very fast and can easily be implemented in the system, but may reduce the speed of the measurement system.

9.2.2   Differential methods

A differential circuit is designed to compare outputs of two parallel circuits. If certain criteria are met, the outputs are identified as noise and cancel each other. Otherwise the outputs are considered as PDs in the equipment and measured accordingly.

9.2.2.1   Balanced circuits

With a balanced circuit connection, PDs inside one item of equipment and noise entering from the terminal will give different responses S1 and S2 at the sensors, as shown in Figure 9.3. Discharges inside C1 or C2 produce different amplitudes and opposite polarities at the sensors, whereas noise from outside the equipment generates pulses of the same polarity, similar waveshape and almost equal amplitudes under favourable conditions. C1 and C2 could be two single-phase transformers, two bushings of a circuit breaker, two HV current transformers or a voltage transformer and a discharge-free capacitor. Although this is a well-known technique, the following points may be ignored, resulting in unsuccessful results.

(a) The measurement system must have enough resolution, which is usually achievable with wideband detectors. A poor resolution can cause overlapping of subsequent pulses, making it difficult to identify signal pairs from the same source.

(b) S1 and S2 caused by terminal noise may have different amplitudes if C1, C2 and associated circuits are not identical in all frequency ranges. Terminal connections may also contribute to the difference. The problem can be alleviated if the two measurement channels can be calibrated by injecting at the common terminal and compensated by adjusting Z1 and Z2.

(c) Although the time delay between S1 and S2 may cause inaccuracies, it can give additional information for discrimination of noise. This delay may be detectable for cables and certain types of transformer or generator windings.

(d) If C1 and C2 are items of equipment with HV terminals connected to long busbars, possible signal coupling between the busbars should be considered. The coupling coefficient varies in different conditions and careful calibration would be necessary.

9.2.2.2   Noise discrimination using additional discharge-free capacitor

Dave Allan of PowerLink, Australia [6], has developed a system to detect online PDs in transformer bushings using an additional discharge-free capacitor, as shown in Figure 9.4. The capacitor can be raised to touch the conductor connected to the terminal of the bushing. Signals are detected from the bushing capacitor tapping and from the PD-free capacitor for comparison. Noise from outside the transformer can then be identified. The technique has been used successfully in several substations up to 275 kV, especially for CTs in which PDs <100 pC were measured.

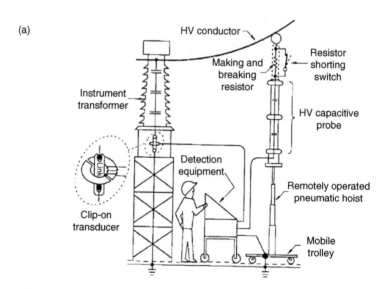

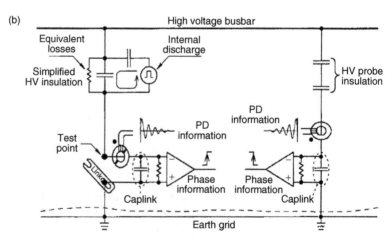

Figure 9.4   Online detection of PDs in a transformer bushing using an additional discharge-free capacitor [6]: (a) pictorial representation of the detection system; (b) principle of the technique [courtesy of Dave Allan]

9.2.2.3 Parallel circuits in HV equipment

By using two HV devices of different phases where high-frequency coupling exists, it is possible to discriminate noise by comparing the polarities of the first peaks of PDs occurring at the same time on each phase, e.g. between transformer bushings and between capacitive instrument transformers. However, in some cases a parallel circuit may be found in one piece of HV equipment for differential PD measurements. A circuit suggested by Malewski [7] is shown in Figure 9.5, where discharge pulses are detected by two high-frequency current transformers clamped around the neutral lead and the tank grounding lead of a transformer. The pulses detected at the neutral lead are reversed and added to the pulse from the tank grounding lead following certain attenuation. The sum (i.e. difference between the two CT outputs) is filtered and measured. The noise from outside the transformer produces pulses of the same polarity and are cancelled in the system, whereas internal PD signals of opposite polarity are enhanced.

Another possible differential circuit consists of two shields preinstalled in a reactor for PD detection. The two brass shields are semicylindrical in shape and located symmetrically inside the tank with leads connecting to the detection terminals. The shields act as capacitive couplers to sense PDs in the winding. Noise entering the reactor at the terminal will show the same polarity of voltages at the detection terminals, whereas

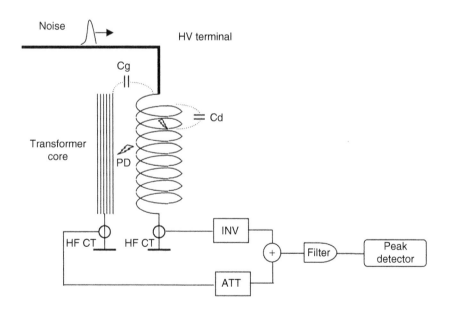

Figure 9.5   *A differential circuit for noise discrimination for PD measurements on power transformers [7], where C_d – capacitance between winding turns and disks, and C_g – capacitance between HV terminal shield, winding and transformer core. Two HF CTs are clamped around the earthing leads of the core and the winding respectively*

(a) Original PD pattern   (c) Class pulse   (d) Sub-pattern   (e) Identification

(b) Classification map

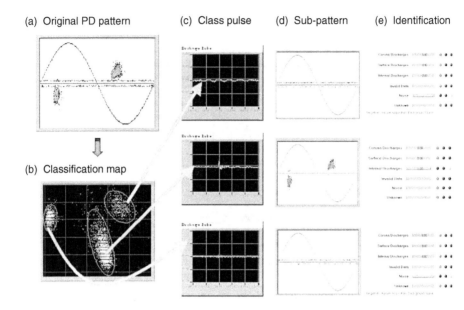

Figure 9.8   PD separation and noise-identification technique developed by TechImp
[9]: (a) original PDs and noise in the phase resolved distribution;
(b) PDs and noise separated to three regions in the equivalent time-
length/equivalent bandwidth plane; (c) typical PD pulses and noise
in the three regions; (d) the phase-resolved distributions of PDs and
noise separated in (b); (e) the type of PD or noise identified by a fuzzy
classification method [courtesy of Prof. G.C. Montanari]

range in which an interleaved winding can be approximated by a capacitive ladder network [10]. In ideal situations, the position of a charge may be uniquely determined by the ratio of the capacitively transmitted pulses at both terminals of the winding. As shown in Figure 9.9, after filtering with a pass-band of 300–400 kHz, which falls within the capacitive range of the model transformer winding, the ratio is monotonous with the injecting position. This suggests a method for locating discharges and discriminating noise from outside of the equipment.

Figure 9.10(a) shows the measurement results on a 66 kV transformer winding presented in three-dimensional form [10,11]. There were two discharge sources in the winding and a corona discharge at the terminal. The ratio of each pair of terminal pulses was used to determine the discharge position and displayed on the x-axis, the discharge magnitude on the y-axis and the number of discharges on the z-axis. It can be seen that the two discharge sources can be easily identified in the 3D graphs and the corona noise identified and removed from the analysis. The PDS and corona are separated according to the ratio of terminal pulse pair and analysed in the format of number-versus-magnitude (pC) and number-versus-phase distributions, as shown in Figure 9.10(b). The type of PD, or otherwise a corona, may be identified from the

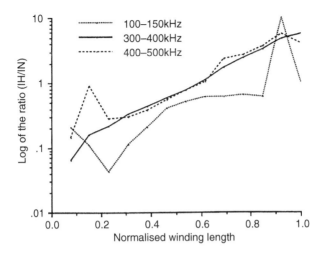

Figure 9.9   *Plot of the ratio of terminal voltages versus the position of simulated discharges injected at various positions along a model transformer interleaved winding after digital filtering with various pass-bands*

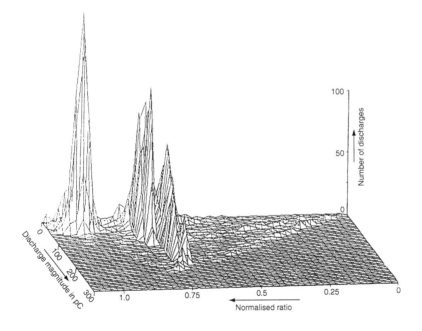

Figure 9.10   *(a) A 3D pattern of PD activities of two discharge sources in a 66 kV interleaved winding and corona at the terminal [11]. x-axis: the ratio of terminal discharge pairs; y-axis: the magnitude in pC; z-axis: the number of discharges.*

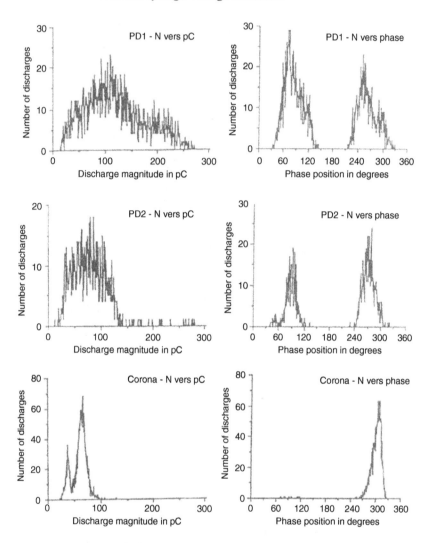

Figure 9.10   *(b)   Based on their ratios of terminal pulse pair, the two PDs and corona are separated and analysed in number-versus-magnitude (pC) and number-versus-phase distributions. The number-versus-phase distribution of corona shows its distinguishing characteristic that the majority of discharges occur in the AC negative half cycle.*

characteristics, for example, the majority of corona discharges are in the AC negative half cycle. If the terminal disturbance includes internal PDs at the line end of the winding, software methods such as waveform analysis can be applied for separating the two types of pulse. These techniques can also be used to separate discharges from the different locations, thus allowing independent analyses. The method is excellent for assisting in the assessment of local insulation problems.

9.3 Non-electrical online condition monitoring

Online condition monitoring is preferable for continuous assessment of equipment integrity and prevention of incipient faults. The more widely accepted online insulation monitoring techniques for HV equipment rely on non-electrical characteristics such as temperature changes, acoustic emission from partial discharges, decomposition of the oil, including increase in gas content, and deterioration of the gas in GIS.

9.3.1 Temperature monitoring of the insulations

Abnormal temperature on equipment surface may indicate a deterioration of insulation losses, an increase of leakage current or a loose contact. The application of optical-fibre sensing techniques for measuring the localized changes of insulation temperatures within transformers and cables (DTS) is discussed in Section 8.2.3. Other temperature-detection methods include thermal scanning and irreversible temperature labels, which are portable or of lower cost.

9.3.1.1 Thermal scanning

Various infrared cameras are available on the market for remote thermal scanning. Because there is no need to install a sensor and the measurement is at a safe distance, the thermal scanners are widely used in the industry. A typical infrared imager is shown in Figure 9.11(a).

9.3.1.2 Irreversible temperature labels

Irreversible temperature recording labels can accurately and economically sense and record surface temperatures. Each label contains one or more sealed temperature-sensitive chemical indicators that change permanently and irreversibly from silver to black at its calibrated temperature. The response time is less than one second and, depending on the temperature rating, an accuracy of ±1 or 2 per cent is achievable. The labels are of miniature size, weight and thickness (nominal 0.01″), which allows installation in areas and on parts that are not practical for other instruments. The label can be useful for temperature monitoring of small transformers, isolators, VT, CT and switches. A photo of a label is shown in Figure 9.12.

9.3.2 Online DGA

Dissolved-gas analysis in oil (DGA) has been widely used for the diagnosis of HV equipment with paper/oil insulation such as transformers, reactors and cable joints for many years and in many cases is the primary method of monitoring. Although periodic sampling and tests in the laboratory (offline test) have been successful in diagnosing incipient faults, there are a number of problems associated with offline DGA testing: for example:

(a) the sampling frequency has to be conservatively selected to be effective; even so, offline periodic tests may fail to predict some failures;

Figure 9.14   Online hydrogen monitoring on a transformer [13] [courtesy of GE Oil & Gas, http://www.gepower.com/prod_serv/products/substation_md/ en/monitoring_instr_sys/hydran.htm]

A particular instrument for continuous monitoring is shown in Figure 9.14. Normally, one or a few gases are monitored by the instrument. The Hydran, made in Canada, is specifically designed to detect hydrogen and carbon monoxide in the oil. Other instruments may detect additional gases, but are more costly.

9.3.3   Acoustic-based techniques for PD detection

9.3.3.1   Transformer windings

The ultrasonic impulses transmitted through the mechanical structure of a winding associated with PDs may be used for location. If it is impossible to utilize electrical pulses from the PD and the physical conditions are favourable a discharge source may be located using a number of appropriately placed acoustic transducers. Reasonable accuracy is obtained if reflections are not significant and only direct path signals are present [14,15]. Triangulation location systems may also be used to pinpoint a location in space, provided it does not lie deep within the transformer or below a certain magnitude. An advantage of these methods is their independence of electrical noise. In some cases the detecting sensitivity has not been satisfactory, especially when the discharge sources are deep inside the coils. Enhancement of the signals may be realized by digital signal processing and averaging. However, if there is more than one discharge source and large interference is present, averaging of the acoustic signals from different sources is likely to result in confusing results. Further work is still being carried out, in particular with respect to improving sensitivity and accuracy, location of inner winding faults and interpretation of characteristic oscillations with both external and internal acoustic transducers.

9.3.3.2   Cable terminations

Acoustic PD measurements have also been developed for cable terminals and proved successful. A particular detector consists of a fibreglass stick or pipe, a sensor and

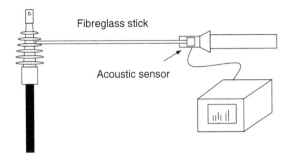

Fibreglass stick

Acoustic sensor

Figure 9.15   *Acoustic PD detector with long fibreglass stick used for voltages up to 33 kV [17]*

a monitor, as shown in Figure 9.15. Since the insulated stick/pipe can be directly applied to the cable terminal and works as a sound-wave guide, the sensitivity of measurement is higher than for other acoustic sensors placed at a distance from the discharge source.

9.3.3.3   Portable acoustic PD detector

Acoustic guns are also used for the detection of discharges around equipment surfaces such as the HV terminals (for corona discharges). As shown in Figure 9.16, the battery-powered portable PD detector can be used for PD detection on HV equipment. With the directional feature, the detector can help to locate PD sources and alleviate the noise problem in site measurements. However, for internal discharges in the insulation, the

Figure 9.16   *A battery-powered detector used in PD detection and location of HV equipment [courtesy of May Elektronik GMBH, http://www.may-elektronik.de/]*

sensitivity may be reduced. Nevertheless, they are used widely in the laboratory and at site to detect corona discharges due to insulation design problems, cracking on porcelain insulators and bad contact, etc.

9.4   Online acoustic/electric PD location methods for transformers

Examples are given of the application of acoustic transducers used in conjunction with (i) the associated electrical pulse measured at the terminal (9.4.1) and (ii) the radiated electrical pulse detected by a sensor within the transformer tank (9.4.2). The viability of the methods depends on knowledge of the velocity of sound though the oil (approximately 1.4m/ms) and the fact that the electrical pulses are detected almost instantaneously.

9.4.1   *Acoustic transducers and winding terminal measurements*

The usual combined tests are carried out with a number of acoustic transducers external to the tank and records obtained at the terminal (bushing tap) of the electrical pulses as well as those from the transducers. By triggering on the acoustic pulses and determining the prior time-delay range expected for the electrical signal – estimated from the tank dimensions and the structural layout – it is possible to obtain an indication of the probable location of a PD source.

An improvement on the above procedure is to measure electrical PDs at both the HV terminal and the neutral and apply the signal separation methods detailed in Section 9.2.4 in order to achieve signal enhancement. This also enables an estimation of the PD magnitude at the source if within the winding. A new technique has been developed using a combined electric/acoustic method [15]. The measurement circuit for a 330 kV single-phase transformer is shown in Figure 9.17. After digital filtering of the discharge signals measured at the HV terminal and neutral, the ratio of each pair of signals was used to separate them according to their different source locations and distinguish each from external interference. The electrical signals that were likely to be occurring at the same discharge source were averaged, as were their associated acoustic signals, resulting in a significant improvement in the detection sensitivity and location estimate as shown in Figure 9.18. By suitable calibration techniques and knowledge of the probable location within the winding, an estimate of the PD magnitude at its site may be possible.

9.4.2   *Application of internal combined acoustic and VHF/UHF transducers*

A particular system for location of PDs in transformers has been developed by the University of Technology, Sydney, and Siemens Ltd [16]. The technique utilizes specially designed composite acoustic transducers, which are matched to the acoustic impedance of the oil and have a short ring-down time. Around each transducer is located a capacitive ring for detection of the radiated RF pulse. The capacitive sensor

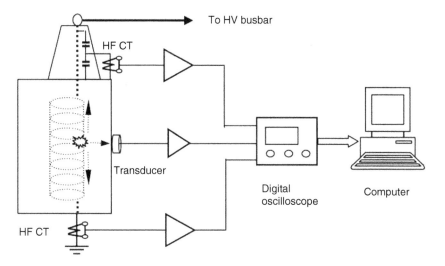

Figure 9.17   *Circuit connection of PD measurement on a 330 kV single-phase transformer using a combined electric/acoustic PD detection method*

operates in a tuned circuit within 10–100 MHz. The combined unit is inserted through the transformer tank side/lid mounted on an inspection plate as shown in Figure 9.19. A modified arrangement allows insertion through an oil drain valve. The face of the combined transducer is kept flush with the inner surface of the tank thus minimizing the possibility of forming a hazard during the transformer operation.

Due to the internal mounting, interference effects are reduced and sensitivity increased compared with externally located transducers. Acoustic matching of the sensor is good. However, the versatility in positioning is reduced. Also, the VHF sensor is not calibrated in terms of pCs. The system can incorporate continuous remote monitoring of any changes in the output of a number of acoustic transducers, which may identify a PD based on correlation of the sound and RF signals. The signal processing includes estimates of the locations in relation to the sensors.

9.5   Electrical online condition monitoring

Although offline insulation tests, either destructive or non-destructive, are very useful for the assessment of HV equipment condition, as indicated in Section 9.1, it is preferable to use online tests. With electrical online tests, the insulation defects can be directly assessed from the values and changes of the electrical properties. However, since all HV equipment is under high electrical stress, the electrical interferences such as corona and harmonic voltages always make online electrical monitoring more difficult. Advanced techniques in hardware design and software analysis have to be utilized in achieving the desired results.

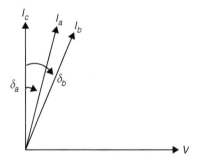

Figure 9.20   Phase difference between voltage and current for a lossless capacitor, and for phases a and b with different losses

Table 9.2   Identification of the faulty phase from comparison – examples

	$\Delta\delta_{ab}$	$\Delta\delta_{bc}$	$\Delta\delta_{ca}$	Remarks
Comparison between phases	Larger	Larger	Small	Phase b may have an insulation problem
Change with time	Small	Larger	Larger	Phase c may have an insulation problem

problem in one phase. By monitoring the difference between phases and its changes with time, the faulty phase can be identified and repaired accordingly, as shown in Table 9.2.

9.5.1.3   Practical online measurement system

A few integrated DLA and capacitance-monitoring systems have been developed in the world for online continuous condition monitoring of CTs, capacitive VTs and transformer bushings. The one developed by Dave Allan of PowerLink, Australia, is shown in Figure 9.21 [6]. The system comprises current (voltage) sensors to measure the current waveform, phase encoders to pick up the phase information from wave-shape conditioning, an optical-fibre transmission system and DLA calculation and display unit. The arrangement has been installed in a number of switchyards.

9.5.2   Online leakage current measurement

Most outdoor HV equipment has porcelain housings, e.g. CTs, VTs, bushings and surge diverters. Due to contamination, moisture or cracking of the porcelain housings, surface creepage discharges/currents may occur, which can significantly reduce the electric strength. Online monitoring of the leakage current can be used to detect

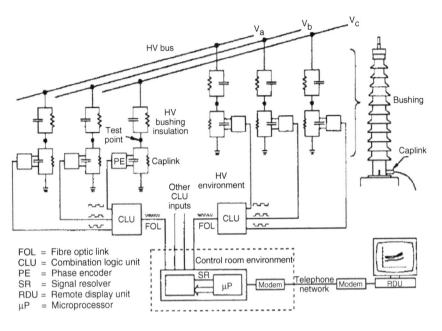

Figure 9.21 *An online DLA monitoring system for CTs, CVTs and transformer bushings in a substation [6] [courtesy of Dave Allan]*

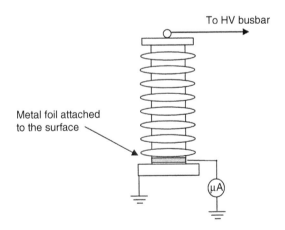

Figure 9.22 *Online leakage current monitoring*

incipient failure. The detecting circuit is depicted in Figure 9.22. If the low-voltage end of the porcelain housing is directly grounded, an additional electrode made of metal foil may be installed around the housing surface to detect the leakage current. Similar systems are used for monitoring high-voltage surge diverters.

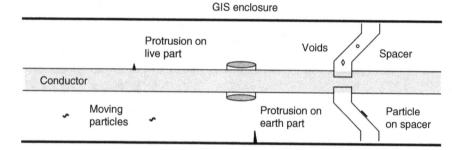

Figure 9.23   The main defects inside GIS resulting in PDs [18]

9.5.3   Electrical online PD detection

9.5.3.1   GIS PD detection

Due to the completely sealed HV circuit and the use of pressurized gas (SF_6) as the main insulation, GIS has the advantage of a compact structure and low maintenance requirements. Although GIS are usually very reliable, online PD detection is necessary for commissioning tests and condition monitoring for detecting any developing defects.

Metallic-particle contaminants are the most critical defects in GIS. They may develop from:

1. mechanical abrasions;
2. movement of conductors under load cycling;
3. vibrations during shipment and in service.

Defects and particles may be created during the production of the unit in the factory or during assembly at site. In normal operation, particles could also be produced by fast-earthing switches or disconnector switches. The main defects in GIS are depicted in Figure 9.23.

Conventional test methods (IEC 60270) are usually applied during commissioning but these tend to be insensitive due to the poor coupling between the sensors and the PD sites. To obtain the required sensitivities of <5 pC, UHF sensors and associated systems, including remote monitoring, have been developed [18]. The frequency band might be up to the GHz range depending on different design and sensitivity requirements. At such high frequencies, each sensor can cover only a certain distance of the compartment/busbar. Therefore multiple sensors have to be installed to monitor the whole GIS. Typical UHF sensors used in GIS are shown in Figure 9.24.

9.5.3.2   Transformer PD detection using VHF and UHF sensors

In addition to online acoustic location of PDs in transformers, as indicated earlier, a number of organizations have developed systems for measuring the PD values on line. These are based on:

1. single-ended measurements using the bushing tap only and advanced signal processing including phase resolved analysis for identification and separation of interference [19];

(a)   (b)

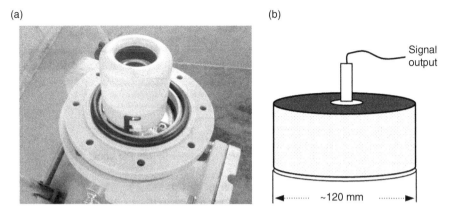

Figure 9.24   *UHF sensors used for PD detection in GIS: (a) the sensor attached to the earthing switch inside GIS; (b) the sensor used externally on the observation window [18]*

2.   multiple-terminal measurements including neutral and tank ground connections and appropriate signal processing to isolate interference effects [20].

Recent techniques aimed at reducing interference problems have applied the knowledge acquired from GIS investigations involving UHF detection methods. The KEMA patented system [21] includes capacitor-type sensors, as does that developed by the University of Strathclyde [22]. The latter studies show that at UHF it should be possible to locate PDs and characterize them in respect of energy and phase position.

9.5.3.3 Online PD detection on power cables

There are mainly two types of power cable – paper/oil and XLPE. These have significantly different PD levels. For XLPE cables, the allowed PD magnitude should be under 5–50 pC depending on the different voltage levels. On the other hand, paper/oil cables can tolerate larger PDs of even a few hundred pC. Therefore, online PD measurements at the terminal could be successful for paper/oil cables if appropriate techniques are used to discriminate against interference. For XLPE cables, more advanced techniques should be used to achieve a higher S/N ratio. Probably the most widely accepted technique for online PD detection is VHF/UHF sensing, which is specifically used for PD detection on cable joints. It has been found that cable joints and terminations are still the weakest part of a cable circuit.

In order to detect PDs in cable joints, prefabricated capacitive sensors imbedded inside the joint may be used, giving higher sensitivity and reducing noise problems. However, these internal sensors tend to make the cable structure more complicated and more expensive.

Sensors can also be installed outside the cable joint, such as the one developed by some Japanese companies [23]. The principle of this technique is shown in Figure 9.25. Two metal-foil electrodes are installed around the cable on each side of the joint. The capacitance between the foil and the conductor forms the coupling

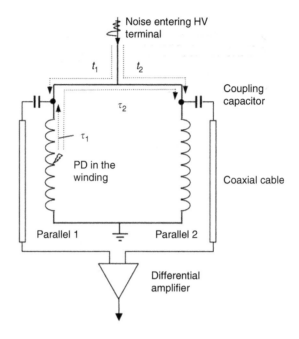

Figure 9.27   Ontario Hydro differential coupler installation [29] (note: for external noise $t_1 = t_2$ and PD in winding $\tau_1 \neq \tau_2$)

The PDA detects the steep pulses, which are considered to be capacitively coupled between the end windings of each turn, as little of the high-frequency components are conducted through the slots. From regular measurements on an individual machine and comparison with the historical data, the changes in discharge performance can be monitored and the insulation condition evaluated from previous knowledge of itself and similar units.

9.5.3.7   Ontario Hydro – IRIS Turbo-Generator PD Analyser (TGA)

Due to the different structural configurations of turbo-generators and larger interference levels, a stator slot coupler for sensing PDs in turbo-generators has been developed [31]. The sensor is essentially a directional electromagnetic coupler fitted in the slot, under the stator winding slot wedges between layers of wedge packing. Because of its location close to the slot conductors, the sensor is sensitive only to partial discharges but not the interference because of the filtering effect of the stator structure. Output voltages of the sensor have very high frequency (rise time being within a few ns), which are measured by a specially designed detector. The layout of the sensor is shown in Figure 9.28. The number of sensors is normally six in this system.

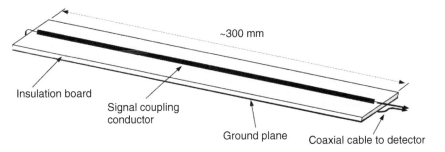

Figure 9.28   The simplified schematic of the stator slot coupler [31]

9.5.3.8   UNSW–SMHEA PD detector

In the late 1980s, a team at the University of New South Wales, in collaboration with engineers from Snowy Mountains Hydro-Electric Authority developed a PD detection method for hydraulic generators [32,33]. The method is based on the travelling wave and capacitively coupled PD signals detected at the HV terminal of the stator winding. The capacitively coupled part of a PD in the winding appears at the terminal almost instantaneously, whereas, the travelling wave part takes up to 7–8 μs to reach the same terminal depending on the PD source location and the stator winding structure and size. The time delay is then measured, which can be used to discriminate PD from noise as well as locate the discharge. The circuit connection is shown in Figure 9.29 and the PDs detected on a 200 MW hydro generator are shown in Figure 9.30.

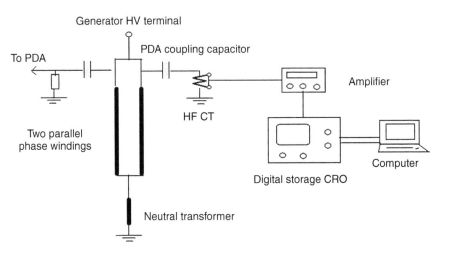

Figure 9.29   Test circuit connection (only one phase winding is depicted)

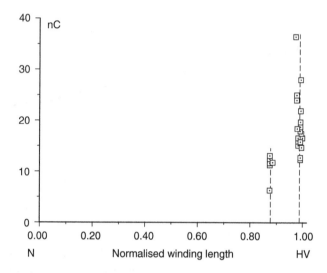

Figure 9.30   *Apparent charge magnitudes versus estimated locations of the PDs detected on a 200 MW generator showing two possible PD sources at 87.4 per cent and 98.5 per cent of the winding [33]*

9.6   Summary

Although condition monitoring on in-service equipment is difficult due to the noise problem, new techniques have been developed that make the operation of HV equipment safer and more reliable. There are two forms of online monitoring: (i) periodical measurements utilizing built-in sensors with portable instrumentation; and (ii) complete systems continuously recording the data. Online continuous condition monitoring should be integrated into a centralized monitoring system, which needs an advanced communication system using optical fibres, digital radio and even infrared techniques. This has placed a challenging task for future condition monitoring engineers who should be well equipped with knowledge in these areas as well as the operation and physical structure of the HV equipment.

9.7   References

1. Konig, G., and Feser, K., 'A new digital filter to reduce periodical noise in partial discharge measurements', presented at the 6th International Symposium on High Voltage Engineering, New Orleans, 28 August–1 September 1989, paper 43.10
2. *GA-1 Gating Amplifier*, instrument manual, Insultest Australia, November 1997
3. Borsi, H., and Hartje, M., 'New methods to reduce the disturbance influences on the in-situ partial discharge measurement and monitoring', presented at the 6th

International Symposium on High Voltage Engineering, New Orleans, August–September 1989, paper 15.10

4. Su, Q., 'Application of digital signal processing techniques for noise suppression in partial discharge measurements', *Proceedings of the 4th International Conference on Properties and Applications of Dielectric Materials*, Brisbane, 3–8 July 1994, pp. 602–5

5. Su, Q., 'An adaptive filtering method for noise suppression in partial discharge measurements', presented at the International Conference on Electrical Insulation and Dielectric Phenomena, Pocono Manor, Pennsylvania, October 1993, pp. 481–6

6. Allan, D.M., Brundell, M.S., and Boyd, K.J., 'New insulation diagnostic and monitoring techniques for HV apparatus', presented at the 3rd International Conference on Properties and Application of Insulation Materials, Tokyo, July 1991, pp. 448–51

7. Malewski, R., Douville, J., and Belanger, G., 'A diagnostic system for in-service transformers', CIGRE 27, Paris, August–September 1986, paper 1201

8. Black, I.A., and Leng, N.K., 'The application of the pulse discrimination system to the measurement of partial discharges under noise condition', presented at the IEEE International Symposium on Electrical Insulation, June 1980, pp. 167–70

9. Montanari, G.C., Cavallini, A., and Puletti, F., 'A new approach to partial discharge testing of HV cable systems', *IEEE Electrical Insulation Magazine*, January/February 2006;**22**(1):14–23

10. Su, Q., and James, R.E., 'Analysis of partial discharge pulse distribution along transformer windings using digital filtering techniques,' *IEE Proc. C, Gener. Transm. Distrib.*, September 1992;**139**(5):402–10

11. James, R.E., Phung, B.T., and Su, Q., 'Application of digital filtering techniques to the determination of partial discharge location in transformers', *IEEE Transactions on Electrical Insulation*, August 1989;**24**(4):657–668

12. Duval M., and dePablo A., 'Interpretation of gas-in-oil analysis using new IEC publication 60599 and IEC TC 10 database', *IEEE Trans. Dielectr. Electr. Insul.*, March/April 2001;**17**(2):31–41

13. Ritchie, D., 'The opportunities presented by online transformer gas monitoring', presented at the Doble Conference, Boston, April 1999

14. Jones, S.L., 'The detection of partial discharges in transformers using computer aided acoustic emission techniques', record of the IEEE International Conference on Electrical Insulation, Toronto, Canada, June 1990

15. Blackburn, T.R., James, R.E., Su, Q., Phung, T., Tychsen, R., and Simpson, J., 'An improved electric/acoustic method for the location of partial discharges in power transformers', *Proceedings of the 3rd International Conference on Properties and Applications of Dielectric Materials*, Tokyo, July 1991, pp. 1132–35

16. Unsworth J., Booth N., Tallis D., and Ball K., 'Evaluation of novel online partial discharge monitor for high voltage transformers during operation.' Presented at the 16th Nordic Symposium on Electrical Insulation, Copenhagen, June 1999, pp. 133–40

Chapter 10
Artificial-intelligence techniques for incipient fault diagnosis and condition assessment

- Database and expert systems for fault diagnosis
- Transformer DGA fuzzy-logic diagnosis
- Fuzzy-logic condition assessments and ranking

In general, fault diagnosis may be defined as a problem of pattern recognition. The pattern vector consisting of detected parameters can be correlated with its counterpart – a set of classes or types of incipient fault and no-fault. From the observed symptoms or/and test results (data) of equipment, an expert could make a general reasoning to derive the cause or causes using his/her knowledge and experience. However, it may be more objective and accurate when artificial intelligence is employed to establish the links between the cause and results behind the observations. An artificial-intelligence technique could consolidate the knowledge and experiences from a number of experts. It could even learn and improve itself from the cases it examines. During the last decade, many applications of artificial intelligence for incipient fault diagnosis of HV equipment have been developed in the world. It is not within the scope of this book to include all different methods and techniques. This chapter concentrates only on some of the techniques and applications developed by the authors. Interested readers can refer to relevant technical papers and books for detailed information.

10.1 Database for condition assessment

The number of HV devices in an electrical utility runs into several thousands. The devices need to be tested periodically for the monitoring of their insulation conditions. Many diagnostic tests can be carried out on each device, which produces an enormous number of data. Without a suitable computer database, manual storage and analysis of the data would be very difficult as well as time-consuming. Also, when a piece of HV equipment is in question, it is often required to provide the management board with a sound and economic basis for the major decisions that will have to be made concerning possible refurbishment, life extension or replacement. Although condition monitoring can give an early warning about the insulation condition, there is no single technique that alone can guarantee to detect all the range of faults and give a reliable estimation of remnant life. For this reason, combined monitoring systems are required to provide a pool of data that will be analysed thoroughly so that an integrated assessment of equipment condition may be achieved. The accurate assessment also depends on the historical test record to establish a trend associated with normal operating conditions.

10.1.1   A computer database and diagnostic program

In most cases, accurate diagnosis of a piece of HV equipment or a cable needs the application of a number of techniques. The analysis of combined diagnostic test results and their trends can be made easier with the aid of a computer database.

A database to meet all the above requirements should have the following features:

* can store and retrieve a large number of data;
* is easy to use and fast in data retrieving;
* can process data and present results in graphical and tabular forms;
* can be incorporated within expert systems and other artificial intelligence;
* is reliable and secure.

The database software package developed by Monash University HVICM Group in the 1990s fulfils most of the above requirements. The front page of the ICMT-2.0 program is shown in Figure 10.1. Some advantages of the database – fuzzy logic – expert system are listed below.

* **Ease of use and large storage**: All options on the screen are in pull-down form with appropriate instructions. Information about all HV equipment in each power station of a utility (rated voltage, current, power and years in service, etc.) can be stored.
* **Graphical presentation**: Trends of test results can be illustrated graphically, which gives a better assessment of the insulation condition. Figure 10.2 shows the trend of dielectric loss angle ($\tan\delta$) of an HV motor presented in graphical form in ICM-2.0.
* **Searching for abnormal test results**: The test results over a certain limit can be determined and displayed on the screen in tabular forms. A number of queries have been designed, e.g. asking to show all transformers with one or more gases exceeding the guided values.
* **Reporting**: The test results can be printed out in a report.
* **Security**: There is a security system. The low-level password is for general users who can only *view* the data. To add to or change the data, a high-level password must be used, which may be issued to authorized personnel only.

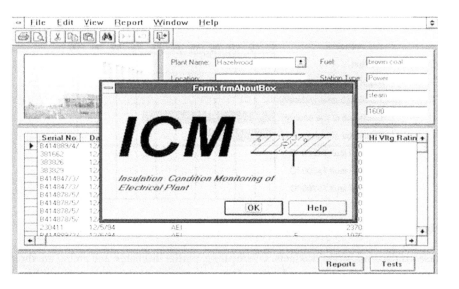

Figure 10.1   The database and expert system ICM-2.0 developed at Monash University

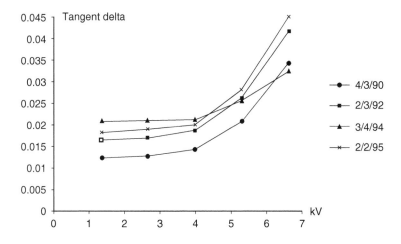

Figure 10.2   The trend of dielectric loss angle (tanδ) of an HV motor against voltage presented in graphical form in ICM-2.0

10.1.2   A combined method for DGA diagnosis

Large power transformers are probably the most important equipment in an electrical transmission system. Correct diagnosis of their incipient faults is vital for safety and reliability of an electrical network. An in-service transformer is subject to electrical and thermal stresses, which can break down the insulating materials and release

Table 10.2   Fault classification according to the IEC Gas Ratio Codes

No.	Fault type	$\dfrac{C_2H_2}{C_2H_4}$	$\dfrac{CH_4}{H_2}$	$\dfrac{C_2H_4}{C_2H_6}$
0	No fault	0	0	0
1	Partial discharges of low energy density	0 (but not significant)	1	0
2	Partial discharges of high energy density	1	1	0
3	Discharges of low energy	1 or 2	0	1 or 2
4	Discharges of high energy	1	0	2
5	Thermal fault of low temperature, $<150\,^\circ$C	0	0	1
6	Thermal fault of low temperature, 150–$300\,^\circ$C	0	2	0
7	Thermal fault of medium temperature, 300–$700\,^\circ$C	0	2	1
8	Thermal fault of high temperature, $>700\,^\circ$C	0	2	2

shown in Table 10.1. In addition, there are also some other limitations in the previous fuzzy diagnosis methods.

In general, in order to diagnose more accurately the incipient faults in a transformer, the key gases should be analysed and the trend of individual faults determined. These could be achieved by the fuzzy-logic method presented in this section. It employs fuzzy boundaries between different IEC codes (i.e. the fuzzy IEC code) with demi-Cauchy distribution function. The critical level of key gases is also treated with the same fuzzy method and combined with the fuzzy IEC codes. Each fault in a transformer can then be assessed by the fuzzy vector and the trend of fault development with time can be closely monitored. For multiple faults in a transformer, this technique can be used to monitor the trend of each fault.

10.2.2.1   Fuzzy IEC codes

According to the IEC codes in Table 10.1, the three gas ratios $r_1 = \frac{C_2H_2}{C_2H_4}$, $r_2 = \frac{CH_4}{H_2}$ and $r_3 = \frac{C_2H_4}{C_2H_6}$ can be coded as 0, 1 and 2 for different ranges of ratios (note, r_1, r_2 and r_3 are larger than or equal to 0). Table 10.1 is rearranged to give a clear relationship between the range of each gas ratio and the corresponding IEC code, as shown in Table 10.3. For example, the codes for a set of gas concentration $\frac{C_2H_2}{C_2H_4} < 0.1$, $\frac{CH_4}{H_2} > 1$ and $1 \leq \frac{C_2H_4}{C_2H_6} \leq 3$ are Code(r_1) = 0, Code(r_2) = 2 and Code(r_3) = 1. According to Table 10.2, the transformer is diagnosed to have a No. 7 fault, i.e. thermal fault of medium temperature 300–700 $^\circ$C.

Table 10.3   Gas ratios and corresponding IEC
codes rearranged from Table 10.1

Ratio $\Leftrightarrow$ Code(r)	0	1	2
$r_1 = \dfrac{C_2H_2}{C_2H_4}$	<0.1	0.1–3	>3
$r_2 = \dfrac{C_2H_4}{H_2}$	0.1–1	<0.1	>1
$r_3 = \dfrac{C_2H_4}{C_2H_6}$	<1	1–3	>3

In the IEC code diagnosis, actually the conventional logic AND and OR are used. For example, the seventh fault is represented by

$$f(7) = code_{zero}(r_1) \text{ AND } code_{two}(r_2) \text{ AND } code_{one}(r_3),$$

where $code_{zero}(r_1)$, $code_{two}(r_2)$ and $code_{one}(r_3)$ are the logics of coded values of gas ratio r_1, r_2 and r_3 respectively. They are either one (true) or zero (false) according to Table 3.

For example, $code_{zero}(r_1) = \begin{cases} 1 & \text{for} \quad r_1 < 0.1 \\ 0 & \text{for} \quad r_1 \geq 0.1 \end{cases}$. Therefore, fault f(7) will be either one (true) or zero (false) by means of the logic operation $code_{zero}(r_1)$ AND $code_{two}(r_2)$ AND $code_{one}(r_3)$.

In the fuzzy-diagnosis method developed, however, the IEC codes 0, 1 and 2 are reconstructed as fuzzy sets ZERO, ONE and TWO. Each gas ratio r can be represented as a fuzzy vector

$$[\mu_{ZERO}(r), \mu_{ONE}(r), \mu_{TWO}(r)],$$

where $\mu_{ZERO}(r)$, $\mu_{ONE}(r)$, $\mu_{TWO}(r)$ are the membership functions of fuzzy code ZERO, ONE and TWO respectively. The membership function is represented by a descending or/and an ascending demi-Cauchy distribution function [11]:

$$\mu_d(r) = \begin{cases} 1 & \text{if} \quad r \leq A \\ \frac{1}{1+(\frac{A-r}{a})^2} & \text{if} \quad r > A \end{cases} \tag{10.1}$$

$$\mu_a(r) = \begin{cases} 1 & \text{if} \quad r \geq A \\ \frac{1}{1+(\frac{A-r}{a})^2} & \text{if} \quad r < A \end{cases} \tag{10.2}$$

where a and A are a pair of parameters that can be selected to give appropriate membership functions [4]. In diagnostic tests, A and a can be regarded as the boundary parameter and distribution parameter respectively. From (10.1) and (10.2), the fuzzy IEC codes ZERO, ONE and TWO are formed. For each set of ratios, the corresponding fuzzy IEC codes can then be determined.

The fuzzy-diagnosis vector $F(i)$ ($i = 0$–8) is then determined by replacing the logic 'AND' with the minimization operation and the logic 'OR' with the

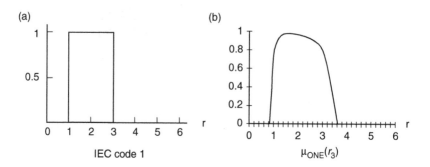

Figure 10.4   Comparison between IEC code 1 and fuzzy IEC code ONE for the gas ratio $r_3 = \frac{C_2H_4}{C_2H_6}$

maximization operation. In other words, the fuzzy multivalue logic is used to substitute for the conventional true-false logic [9]. Based on the IEC rules, different faults (from No. 0 to 8 in Table 10.2) can be diagnosed by the following equations:

$$F(0) = \min[\mu_{ZERO}(r_1), \mu_{ZERO}(r_2), \mu_{ZERO}(r_3)]$$

$$F(1) = \min[\mu_{ZERO}(r_1), \mu_{ONE}(r_2), \mu_{ZERO}(r_3)]$$

$$F(2) = \min[\mu_{ONE}(r_1), \mu_{ONE}(r_2), \mu_{ZERO}(r_3)]$$

$$F(4) = \min[\mu_{ONE}(r_1), \mu_{ZERO}(r_2), \mu_{TWO}(r_3)]$$

$$F(5) = \min[\mu_{ZERO}(r_1), \mu_{ZERO}(r_2), \mu_{ONE}(r_3)]$$

$$F(6) = \min[\mu_{ZERO}(r_1), \mu_{TWO}(r_2), \mu_{ZERO}(r_3)]$$

$$F(7) = \min[\mu_{ZERO}(r_1), \mu_{TWO}(r_2), \mu_{ONE}(r_3)]$$

$$F(8) = \min[\mu_{ZERO}(r_1), \mu_{TWO}(r_2), \mu_{TWO}(r_3)]$$

$$F(3) = \max\{\min[\mu_{ONE}(r_1), \mu_{ZERO}(r_2), \mu_{ONE}(r_3)],$$
$$\min[\mu_{ONE}(r_1), \mu_{ZERO}(r_2), \mu_{TWO}(r_3)],$$
$$\min[\mu_{TWO}(r_1), \mu_{ZERO}(r_2), \mu_{ONE}(r_3)],$$
$$\min[\mu_{TWO}(r_1), \mu_{ZERO}(r_2), \mu_{TWO}(r_3)]\}$$

The equations are then normalized as:

$$F_r(i) = \frac{F(i)}{\sum\limits_{j=0}^{8} F(j)}, \ i = 0\text{--}8 \tag{10.3}$$

A comparison between the conventional IEC code 1 $code_{one}(r_3)$ and the fuzzy membership function $\mu_{ONE}(r_3)$ for gas ratio $r_3 = \frac{C_2H_4}{C_2H_6}$ is shown in Figure 10.4.

10.2.2.2 Fuzzy key gases

Three key gases are commonly used to identify particular faults according to the following rules:

C_2H_2 — Arcing (discharge of high energy)

H_2 — Corona (partial discharge)

C_2H_2 — High temperature oil breakdown

(thermal fault of high temperature > 700 °C)

Each fault is indicated by the excessive generation of the relevant gas above its designated threshold. For different voltage levels, winding structures and types of transformer, the threshold may be different. The threshold can be determined from previous experience for a certain type of transformer. From fuzzy logic theories [9], the membership functions of the fuzzy set 'LOW', 'MEDIUM' and 'HIGH' for each key-gas can then be represented as descending demi-Cauchy distribution function $\mu_L(x)$ or ascending demi-Cauchy distribution function $\mu_H(x)$ or their combination $\mu_M(x)$. For every given key gas concentration x, the corresponding fuzzy vector is therefore $[\mu_{LOW}(x), \mu_{MED}(x), \mu_{HIGH}(x)]$.

The fuzzy-diagnosis vector $F_k(i)(i = 0\text{--}8)$ is then determined in the same way as the fuzzy IEC code.

10.2.2.3 The fuzzy IEC code – key gas method

The fuzzy IEC code – key gas method (FIK) – developed is a combination of fuzzy diagnoses using the IEC codes and key gases. The combined fuzzy-diagnostic vector is represented by

$$F(i) = w_1 F_r(i) + w_2 F_k(i) \quad i = 0\text{--}8 \tag{10.4}$$

where w_1 and w_2 are the weights to relate the fault with the fuzzy IEC codes and key gases respectively. w_1 and w_2 can be 0.5/0.5 or determined from previous experience. $F(i)$ is normalized to make the total fuzzy component equal to 1.

10.2.2.4 Applications

Using the FIK method, a number of 110–330 kV power transformers were diagnosed and some typical results are given in Table 10.4. It can be seen from sample No. 1 that the new method is generally in agreement with the IEC method for transformers of a single or a dominant fault. Compared with the IEC method, the FIK method also has some advantages. For example, due to no matching codes, 13 transformers could not be diagnosed by the IEC method but are diagnosed by the FIK method, as shown in Table 10.4, Nos. 2–3, for some typical results. In some cases, the faults may be only at the early stage or intermittent, which did not produce sufficient gases to give a stronger indication, such as F(2) in No. 2 and F(6–8) in No. 3. However, the information obtained should be useful for future trend analysis. Transformer No. 4 was diagnosed by the IEC method to have a thermal fault of medium temperature

Table 10.5   Gas concentration for the transformers listed in Table 10.4 (in ppm)

No.	H_2	CH_4	C_2H_2	C_2H_4	C_2H_6	IEC ratio codes
1	95	110	<0.1	50	160	0 2 0
2	120	17	4	23	32	1 0 0
3	300	490	95	360	180	1 2 1
4	200	700	1	740	250	0 2 1
5	25	3	0.1	0.1	0.1	below guide level

(300–700 °C). In comparison, the FIK method indicates that both high (>700 °C) and medium temperature (300–700 °C) faults existed. The likelihood of each fault is given by the fuzzy components of 0.477 and 0.431 respectively. The analysis results of another transformer (No. 5) show that, although the gas level is below the guide value, the FIK method can still be used and a low-energy discharge is diagnosed. The fuzzy vector ranges from 0 to 0.441, which could be useful for future trend analysis when the gas level increases.

The gas concentrations for the samples in Table 10.4 are given in Table 10.5. Most transformers have shown a medium or low level of gases and lower increasing rates, such as transformers No. 1, No. 2 and No. 5. Therefore they are closely monitored. Transformer No. 3 was dismantled and an arc damage in the insulation was found in the core. In the inner inspection of another transformer, No. 4, two locations were identified with high-temperature damage due to eddy currents and a bad contact. More transformers can be investigated when certain criteria for individual gases are met. Laboratory tests can also be carried out to fine-tune the fuzzy-diagnosis technique.

The details of the transformers and the DGA results are stored in the database and fuzzy-logic software package, as shown in Figure 10.5. After a transformer is selected, by pressing a button 'Fuzzy Logic Diagnosis', the likelihood of fuzzy elements will be displayed on the screen giving an advice on the predicted incipient fault. Using this computer analysis tool, a number of 110–330 kV power transformers in Victoria, Australia, were diagnosed and some typical results are given in Figure 10.6. For the new transformer, the fuzzy-logic method indicated an 87 per cent diagnosis index for no fault, as shown by Figure 10.6(a), whereas two faults were diagnosed for the second transformer (Figure 10.6(b)). The transformer was eventually untanked and two faults were found in the winding.

The combined DGA diagnosis method (Section 10.1.2) is found useful because it uses three different methods and the last three test results, which involves the trend of faults. When there is more than one fault in a transformer, the fuzzy-logic method may be more effective. The methods have been successfully used for the diagnosis of a number of transformers in Australia. It has been proved that, using the fuzzy-diagnosis method, more detailed information about the faults inside a transformer can be obtained. This is an improvement over the conventional IEC code method, which

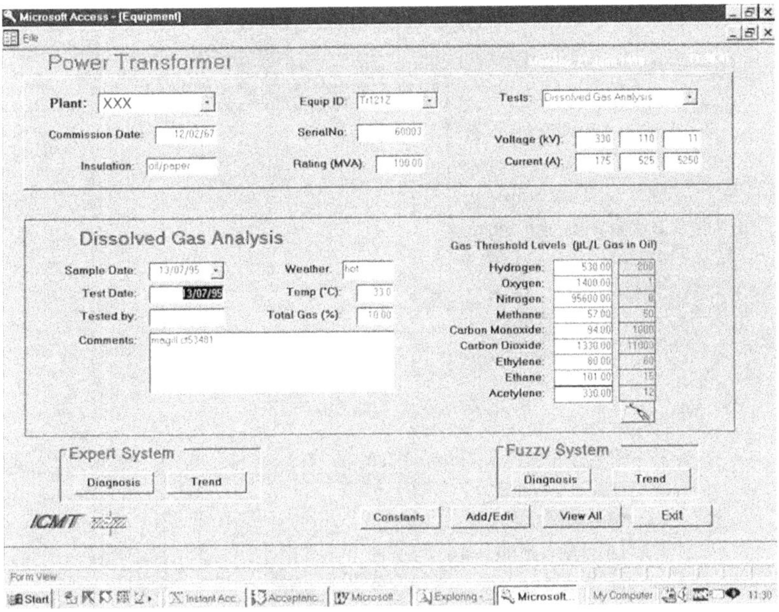

Figure 10.5   The computer program ICM-2.0 works as a comprehensive database and a fuzzy-logic diagnosis tool for transformers

may be due to the more realistic representation of the relationship between faults and dissolved-gas ratios with the fuzzy membership functions. Also, using this method, multiple faults can be diagnosed and their trends determined, which is an important feature.

10.2.2.5   Trend analysis for individual faults

In FIK diagnosis, a fault can be more accurately determined by its fuzzy component, which indicates the likelihood or dominance of the fault. Deterioration of the fault may therefore be closely monitored through trend analysis. This technique has been used for a transformer that was tested over a 15-month period. Thermal faults of medium and high temperature (300–700 °C and >700 °C) were diagnosed by the FIK method and their fuzzy components against the test time are plotted in Figure 10.7. The graph clearly shows the development of each thermal fault in this transformer. It can be seen that at the beginning of this monitoring period, the medium temperature thermal fault F(7) was the main problem of this transformer and the fuzzy component of high-temperature thermal fault (>700 °C) was very small, i.e. below 0.05. The high-temperature thermal fault F(8) was diagnosed from Day 114 onwards and then became stable until Day 406, when the oil was degassed. After degassing, because the thermal faults remained, the fuzzy components F(7) and F(8) went up again from Day 453. It took a few weeks for the gases to be released and dissolved in the oil to

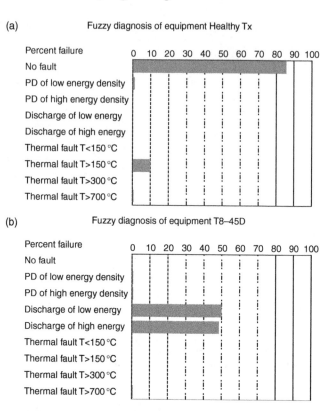

Figure 10.6   Fuzzy-logic diagnosis for transformers using ICM-2.0: (a) a new 'healthy' transformer; (b) a transformer with two types of incipient fault

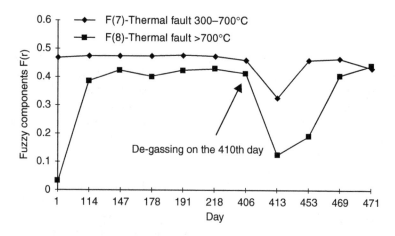

Figure 10.7   The trend of two types of thermal fault in a 330 kV transformer determined by the FIK method

a sufficient level for accurate diagnosis. A small fluctuation of F(8) was recorded on Day 178, which might be due to the lighter load during the specific time period.

It must be noted that, if a transformer has no fault, the fuzzy component F(0) always gives a large value in the range of 0.6–1. For example, the DGA results for a healthy transformer are (in ppm) H_2 – 95, N_2 – 73 000, O_2 – 11 000, CO – 1 000, CH_4 – 20, CO_2–8400, C_2H_4 – 25, C_2H_6 – 45 and C_2H_2 – 2. The fuzzy component of no-fault F(0) = 0.863, which indicates that no fault exists in the transformer. The IEC codes are 0, 0, 0, also indicating no fault. From our experience, when the value of F(0) is between 0.3 and 0.6, an incipient fault at its earlier stage may have occurred. When the fault is getting worse, F(0) will decrease to <0.1.

10.3 Asset analysis and condition ranking

Under the current economical climate, no utility could afford to replace all aged equipment according to its designed lifetime. Priority has to be given to those that have already reached or are close to their critical point of failure. A thorough study of the asset condition and correct ranking of the equipment would be essential for the decision making. This will ensure that the worst equipment can be replaced or refurbished in time.

10.3.1 Equipment ranking according to the insulation condition

Conditions of a piece of equipment can be assessed in terms of its failure probability. Preferably, the whole history of the equipment since commissioning can be retrieved and analysed thoroughly. The following are some important factors to be considered in the equipment ranking:

- failure rate at different ages for a particular type of equipment;
- trend and deterioration rate indicated by diagnostic test results;
- severity of failure consequence;
- value and maintenance costs of equipment;
- working environment, loading and impact of electrical, mechanical and chemical overstresses.

From a combined evaluation of various factors with certain weights attached, the priority list for a type of equipment can be determined.

10.3.2 Insulation health index

The insulation health or failure index may be determined from various test results, quality of equipment, operating condition (e.g. temperature), load, overvoltage impacts and so on. Because there are too many factors that affect the probability of insulation failure, it may not be possible to include every factor in the evaluation. Starting from some important factors, an evaluation system using fuzzy logic has been developed [12–15].

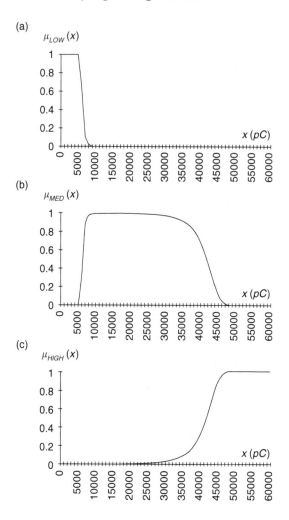

Figure 10.8   The membership functions of fuzzy sets of P_a (LOW, MEDIUM, HIGH)

10.3.4   Example of fuzzy logic condition ranking

In order to check the failure index analysis using the proposed fuzzy-logic technique, 14 hypothetical HV motors and their test results are used, as shown in Table 10.7.

(1) P_a, P_r AND THEIR FUZZY SETS

The level of partial discharge (P_a), its trend (P_r) and their fuzzy sets LOW, MEDIUM, HIGH are calculated as shown in Table 10.8.

(2) T_a, T_r AND THEIR FUZZY SETS

The level of dielectric dissipation factor tan δ (T_a), its trend (T_r) and their fuzzy sets LOW, MEDIUM, HIGH are calculated as shown in Table 10.9.

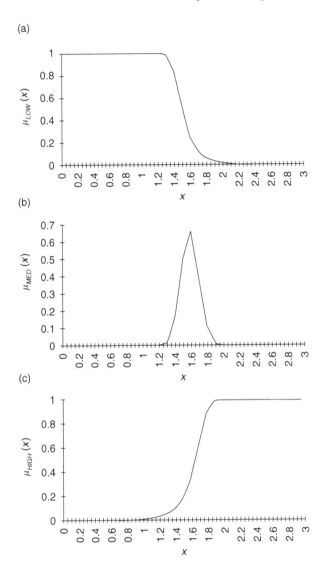

Figure 10.9   The membership function of fuzzy sets of P_r (LOW, MEDIUM, HIGH)

(3) K_a, K_r AND THEIR FUZZY SETS

The level of knee-point voltage (K_a), its trend (K_r) and their fuzzy sets LOW, MEDIUM, HIGH are calculated as shown in Table 10.10.

Using the fuzzy sets of P_a, P_r, T_a, T_r, K_a, K_r and specified weights, the 14 hypothetical motors are examined to give their failure indexes, as given in Figure 10.10. The weights may be selected from experience and can be tuned up in the analysis. It can be seen that motors No. 8 and No. 12 have the highest failure indices and should

Table 10.7   Six important parameters of the hypothetical HV motors

No.	P_a (pC)	P_r	T_a	T_r	K_a	K_r
1	3 000	1.4	0.01	1.3	0.8	0.2
2	5 000	1.3	0.015	1.4	0.75	0.15
3	8 000	1.35	0.012	1.2	0.75	0.5
4	10 000	1.4	0.015	1.3	0.7	0.65
5	20 000	1.6	0.02	1.1	0.7	0.5
6	25 000	1.8	0.01	1.2	0.7	0.5
7	10 000	1.3	0.008	1.6	0.7	0.75
8	30 000	1.3	0.015	2	0.65	0.9
9	40 000	1.5	0.01	1.2	0.65	0.5
10	50 000	1.4	0.03	1.4	0.65	0.75
11	60 000	1.3	0.02	1.3	0.7	0.6
12	10 000	2	0.005	1.1	0.7	0.9
13	35 000	1.1	0.015	1.2	0.65	0.5
14	80 000	1.4	0.012	1.1	0.65	0.8

Table 10.8 P_a, P_r and their fuzzy sets of the hypothetical HV motors

No.	P_a (pC)	LOW	MED.	HIGH	P_r	LOW	MED.	HIGH
1	3 000	0.999	0	0.001	1.4	0.764	0.151	0.085
2	5 000	0.999	0	0.001	1.3	0.938	0.012	0.05
3	8 000	0.025	0.974	0.002	1.35	0.88	0.055	0.065
4	10 000	0.003	0.995	0.002	1.4	0.764	0.151	0.085
5	20 000	0	0.994	0.006	1.6	0.194	0.53	0.277
6	25 000	0	0.988	0.012	1.8	0.056	0.101	0.843
7	10 000	0.003	0.995	0.002	1.3	0.938	0.012	0.05
8	30 000	0	0.971	0.029	1.3	0.938	0.012	0.05
9	40 000	0	0.678	0.322	1.5	0.425	0.425	0.15
10	50 000	0	0	1	1.4	0.764	0.151	0.085
11	60 000	0	0	1	1.3	0.938	0.012	0.05
12	10 000	0.003	0.995	0.002	2	0.019	0	0.981
13	35 000	0	0.914	0.086	1.1	0.98	0	0.02
14	80 000	0	0	1	1.4	0.764	0.151	0.085

be put on the first priority for maintenance or replacement. These two motors have the highest change rate of PD level, DDF or knee-point voltage, which indicates that their rate of insulation condition deterioration is high. Although motor No. 14 has

Table 10.9 T_a, T_r and their fuzzy sets of the hypothetical HV motors

No.	T_a (pC)	LOW	MED.	HIGH	T_r	LOW	MED.	HIGH
1	0.01	0.998	0	0.002	1.3	0.938	0.012	0.05
2	0.015	0.059	0.938	0.003	1.4	0.764	0.151	0.085
3	0.012	0.708	0.29	0.002	1.2	0.969	0	0.031
4	0.015	0.059	0.938	0.003	1.3	0.938	0.012	0.05
5	0.02	0.004	0.99	0.006	1.1	0.98	0	0.02
6	0.01	0.998	0	0.002	1.2	0.969	0	0.031
7	0.008	0.998	0	0.002	1.6	0.194	0.53	0.277
8	0.015	0.059	0.938	0.003	2	0.019	0	0.981 ·
9	0.01	0.998	0	0.002	1.2	0.969	0	0.031
10	0.03	0	0.971	0.029	1.4	0.764	0.151	0.085
11	0.02	0.004	0.99	0.006	1.3	0.938	0.012	0.05
12	0.005	0.999	0	0.001	1.1	0.98	0	0.02
13	0.015	0.059	0.938	0.003	1.2	0.969	0	0.031
14	0.012	0.708	0.29	0.002	1.1	0.98	0	0.02

Table 10.10 K_a, K_r and their fuzzy sets of the hypothetical HV motors

No.	K_a (pC)	LOW	MED.	HIGH	K_r	LOW	MED.	HIGH
1	0.2	0.994	0	0.006	0.2	0.004	0.994	0.002
2	0.25	0.494	0.494	0.012	0.15	0.059	0.94	0.002
3	0.25	0.494	0.494	0.012	0.5	0	0.987	0.013
4	0.3	0.057	0.915	0.028	0.65	0	0.947	0.053
5	0.3	0.057	0.915	0.028	0.5	0	0.987	0.013
6	0.3	0.057	0.915	0.028	0.5	0	0.987	0.013
7	0.3	0.057	0.915	0.028	0.75	0	0.824	0.176
8	0.35	0.012	0.903	0.085	0.9	0	0.107	0.893
9	0.35	0.012	0.903	0.085	0.5	0	0.987	0.013
10	0.35	0.012	0.903	0.085	0.75	0	0.824	0.176
11	0.3	0.057	0.915	0.028	0.6	0	0.968	0.032
12	0.3	0.057	0.915	0.028	0.9	0	0.107	0.893
13	0.35	0.012	0.903	0.085	0.5	0	0.987	0.013
14	0.35	0.012	0.903	0.085	0.8	0	0.657	0.343

the largest PD level and low knee-point voltage, because its PD level is stable and knee-point change rate is not the largest, it is not ranked to the worst motor.

From this example, it can be seen that visual examination of the test results may not be sufficient to determine the condition ranking of a type of equipment. Using fuzzy-logic analysis and suitable weights in the calculation, the failure index could be more accurate in terms of the total condition assessment. Of course, the coefficients

14. Su, Q., 'Reliability centered maintenance of electrical plant – some important issues', presented at AUPEC'97, UNSW, Sydney, September–October 1997, pp. 615–20

15. Su, Q., 'Insulation condition assessment of large generators', presented at the International Power and Energy Conference, Melbourne, 1999, pp. 123–8

10.6 Problems

1. Based on the fuzzy-logic method explained in this chapter, derive the IEC ratio codes and calculate the fuzzy logic elements F(0)...F(7) for the transformers whose DGA test results are given in the following table. Fill in the derived and calculated results in the following tables.

No.	H_2	CH_4	C_2H_2	C_2H_4	C_2H_6	IEC ratio codes
1	95	110	<0.1	50	160	
2	120	17	4	23	32	
3	300	490	95	360	180	

No.	F0	F1	F2	F3	F4	F5	F6	F7
1								
2								
3								

2. Referring to technical papers and books, find the difference between the IEC code, Rogers and triangle methods. Is there any relationship between these methods in terms of the comparison between different gas concentrations?

Appendix 1

Abbreviations

CIGRE   Conseil International des Grands Reseaux Electriques (International Council on Large Electric Systems, Paris, France)

DEIS   IEEE Dielectrics and Electrical Insulation Society

EPRI   Electric Power Research Institute (Palo Alto, California)

ERA   ERA Technology Ltd (Leatherhead, Surrey, UK)

IET   Institution of Engineering and Technology (incorporating the former Institution of Electrical Engineers, UK)

IREQ   Institut de Recherche d'Hydro Quebec (Varennes, Quebec)

ISH   International Conference on High Voltage Engineering (bi-annual conference)

Appendix 2
Major standards organizations

ANSI	American National Standards Institute
ASTM	American Society for Testing of Materials (ASTM Internation
BSI	British Standards Institute
CISPR	International Special Committee on Radio Interference (subcommittee of IEC)
DKE (DIN/VDE)	German Commission of Electrical, Electronic and Informat Technologies (standards and regulations)
IEC	International Electrotechnical Commission
IEEE	Institute of Electrical and Electronic Engineers (USA)
ISO	International Organization for Standardization
SA	Standards Australia (SAI GLOBAL)
WSSN	World Standards Services Network (www.wssn.net/WS index.html)

Appendix 3
Answers to problems

Chapter 1

No calculations.

Chapter 2

Q.1: See Chapter 6.

Q.2: (a) BDV 22.4 kV (RMS) $E_{air} = 2.12$ kV/mm (RMS) $E_{resin} = 0.6$ kV/mm (RMS).

(b) BDV 74.75 kV (RMS) $E_{gas} = 7.07$ kV/mm (RMS) $E_{resin} = 2.02$ kV/mm (RMS).

(c) BDV 84.3 kV (RMS) $E_{oil} = 7.5$ kV/mm (RMS) $E_{resin} = 4.72$ kV/mm (RMS). See Figures 2.2 and 2.3.

Q.3: $C_b = 0.0072$ pF $C_c = 0.0625$ pF $C_a \gg C_b$ $q_c = 285$ pC $q_a \approx 32$ pC $V_i = 31.1$ kV (RMS).

Chapter 3

Q.1: Refer to Chapters 2, 4 and 5.

Q.2 Refer to Chapters 2, 4, 5 and 9.

Q.3: (i) $E_{oil} = 11.8$ kV/mm (RMS) $E_{pbd} = 5.9$ kV/mm (RMS).

(ii) $E_{oil} = 2.7$ kV/mm $E_{pbd} = 27$ kV/mm. Note higher stress in solid.

Q.4: $C_{oil} = 27.8$ pF $C_{pbd} = 129$ pF –Total C $= 22.9$ pF.

$R_{oil} = 7 \times 10^{10}\,\Omega$ $R_{pbd} = 30 \times 10^{10}\,\Omega$.

Q.5: (i) 94 kV (RMS).

(ii) 127 kV (RMS). Reduced oil duct thickness allows higher V_i.

Index

Printed in the USA
CPSIA information can be obtained
at www.ICGtesting.com
JSHW011518221024
72172JS00008B/65